DJ Skills

DJ Skills
The Essential Guide to Mixing and Scratching

Stephen Webber

with Alexia Rosari, Juliette Siegfried, Kate Schutt,
Chad Seay and Daniel Birczynski

Interview transcriptions by Alexia Rosari, Chris Roberts,
Matt Carter, Alex Turkovic and Miles Walker

AMSTERDAM · BOSTON · HEIDLEBERG · LONDON · NEW YORK · OXFORD
PARIS · SAN DIEGO · SAN FRANCISCO · SINGAPORE · SYDNEY · TOKYO

Focal Press is an imprint of Elsevier

Senior Acquisitions Editor: Catharine Steers
Publishing Services Manager: George Morrison
Senior Project Manager: Paul Gottehrer
Editorial Assistant: David Bowers
Marketing Manager: Marcel Koppes
Cover Design: Alan Studholme
Interior Design: Leslie Haimes

Focal Press is an imprint of Elsevier
30 Corporate Drive, Suite 400, Burlington, MA 01803, USA
Linacre House, Jordan Hill, Oxford OX2 8DP, UK

Photos by Stephen Webber.
Edison photos by Rob Jaczko.
Additional photographic assistance and editing by Aubrey Webber, Angela Webber, and Jason Dawson.
DJs in the skills photos: Raydar Ellis, DJ JD, DJ Wrath, and DJ Needlejuice.

 Recognizing the importance of preserving what has been written, Elsevier prints its books on acid-free paper whenever possible.

Library of Congress Cataloging-in-Publication Data
Application submitted

British Library Cataloguing-in-Publication Data
A catalogue record for this book is available from the British Library.

ISBN: 978-0-240-52069-8
ISBN: 978-0-240-52104-6 (CD-ROM)

For information on all Focal Press publications
visit our website at www.books.elsevier.com

Typeset by Charon Tec Ltd (A Macmillan Company), Chennai, India
www.charontec.com

07 08 09 10 11 5 4 3 2 1

Printed in the United States of America

To Jam Master Jay
Now God has a DJ

CONTENTS

ACKNOWLEDGEMENTS

There are several people whom I need to acknowledge for their valuable contributions to this project.

Thanks to Alexia Rosari for her excellent work on the history of records, remix tools, and her invaluable help with the interviews. I appreciate your help more than you'll ever know, and I can't wait to read your books; you are going to be an excellent author.

Thanks to Doc Maji, Jim Tremayne, Ron Principle, Chris Csikszentmihalyi, and Chris Roman for sharing their wonderful insights.

Thanks also to all of the DJs who gave so generously of their time; Logic, QBert, Rob Swift, Shadow, Paul Oakenfold, Kid Koala, BT, Radar, Faust, Shortee, Grandwizard Theodore and Grandmixer DXT. You guys are changing the world.

Thanks to Jason Petrin for production help with the musical examples.

Thanks to my amazing students and colleagues at Berklee who always lend their support; Pat Pattison, Mark Wessel, Rob Jaczko, Mitch Benoff, Dan Thompson, Jeanine Cowan, David Mash, Watson Reid, Roger Brown, and Carl Beatty.

Thanks especially to my wife Susan, and daughters Aubrey and Angela for putting up with the constant work and travel that it took to produce this book.

INTRODUCTION

A month before we lost pioneering Hip-hop DJ Jam Master Jay, the two of us sat down together backstage at the Roxy Theatre in New York to talk about the importance of the DJ in popular culture.

"DJs make the world go 'round," he insisted. While to some this might seem an over-statement or false bravado, Jay really meant it.

A number of enthusiasts trace the lineage of the DJ all the way back to man's earliest tribal dances. Around the camp fire thousands of years ago, someone was taking charge of things; setting the tempo, beating the hollow log groove into overdrive, calling the chants, whipping the dancers into a frenzy. To many, the DJ is the current incarnation of this lord of the dance; ravers in the 1990s even coined the term *techno shaman* to describe this role. In numerous cultures and sub-cultures around the world, dance has once again become the central expression of community, and this time it's the DJ who is setting the tempo.

Today's DJs take many forms: radio DJs wake us up in the morning and put us to sleep at night. Mobile DJs are there to guide us through the celebration of life's central rituals, whether it's coming of age, marriage, anniversary, or retirement. Mobile DJs also ease us into our clumsy mating rituals at school dances, while club DJs take over as the principle facilitators as we become more spirited in nightclubs and raves.

Like X-treme sports figures, scratch and battle DJs expand the boundaries of human possibilities. Selector and remix DJs filter the world we live in and serve it back to us in ways that make us perceive our environment differently. The most popular DJs are millionaire rock stars, adored by the masses and idolized by youth from every corner of the globe.

On the other end of the spectrum, DJing by its very nature lends itself as a postmodern form of artistic expression. Serious artists from the electro-acoustic music world have been exploring the art form for years, and enterprising technical wizards are inventing ways to scratch and otherwise interact with video as well as audio. Christian Marclay, a visual artist, musician and DJ, applies the vocabulary and techniques of the Hip-hop DJ to the aesthetics of modern art. Marclay's 2002 work, Video Quartet, is a virtuosic four screen audiovisual composition, which combines the expertise of the DJ and visual artist in a seamless fashion (Figure I.1).

By studying music history and music technology, one might have predicted that the advent of the record would bring about the creation of the modern DJ, as the advent of musical notation brought about the serious composer and conductor.

At first, musical notation was used as a way of preserving melodies. As musical scribes grew into composers in their own right, DJs have evolved into creators of not only records, but of entire musical movements.

Radio DJ Alan Freed helped create rhythm and blues, then facilitated its transformation into rock 'n' roll by popularizing local styles and bringing them together.

In Jamaica, independent DJs built up massive sound systems, developed a style of *toasting* which urbanized into rap, and made staid corporate-run radio stations irrelevant by championing rock steady, ska and reggae and inventing dub music. It was a transplanted Jamaican DJ by the name of Kool Herc who sewed the seeds of Hip-hop in the South Bronx, and pioneering DJs Herc, Grandmaster Flash and Afrika Bambaataa are Hip-hop's unquestioned architects.

It is now nothing new for entire genres of dance music, including house, breakbeat, jungle, and countless offshoot sub-genres to be conceived of, created, and championed entirely by DJs.

Today it's common for DJs, with their years of experience controlling the dance floor, to be called upon by major record labels to produce popular music's biggest stars, and to remix hit records to give them maximum impact in dance clubs and raves across the planet.

Fig. I.1. DJing with video.

DJs are sought out to become music supervisors and composers for major motion pictures in order to heighten the film's cultural relevance and emotional impact.

DJs are having a profound impact on the musical products industry as well. Hardware manufacturers are rushing to develop tools that will find favor with DJs, from digital turntables to all-in-one groove boxes. New DJ-centric software is cropping up at an astonishing rate, and well-established programs are incorporating features designed to appeal specifically to DJs and remixers.

While often the DJ *is* the band (to the consternation of many musicians), DJs have also taken their place *in* the band. Musical acts as varied as Portishead, Incubus, Slip Knot, Sugar Ray, Kid Rock, Beck, Liquid Soul, Herbie Hancock, Moby, and Madonna have welcomed DJs into their ranks.

The DJ's main tool of expression, the turntable, has become a musical instrument in its own right; the practice of using it as such has been dubbed "turntablism," and its practitioners "turntablists."

For musicians who think (or hope) that all of this is a novelty or a passing fad, consider that as of this writing, Rob Swift is playing the Blue Note with jazz pianist Bob James, experienced bandleader DJ Logic is touring with Bassist Christian McBride (sans drummer!), and DJ Radar is working with the KRONOS String Quartet during a break from recording his turntable concerto, which premiered at Carnegie Hall with full orchestra.

The majority of successful DJs I've come to know share one important trait: musicianship. A surprising number are multi-instrumentalists who were drawn to DJing due to a passion for the art form in its own right, rather than a lack of "traditional" musical ability.

"I was a drummer, then I saw Grandmaster Flash," Jam Master Jay told me. "That changed my whole perspective."

Jazz legend Herbie Hancock enthuses about his turntablists, "DJ Disk, his whole family is musical. They're all musicians. He has a very unusual and advanced rhythmic perspective."

"Grandmixer DXT's not only a turntablist," continues Hancock, "he plays drums, he plays keyboards, he plays bass. He comes from the heart and the perspective of being a musician in the traditional sense."

When electronic musician par excellence and Berklee alum BT refused to accept the moniker of DJ for years, it wasn't because he felt the title was beneath him.

"I didn't want to insult my friends who really *are* world class DJs," he explains.

Like the terms "musician," "artist," or "entertainer," the term "DJ" (originally short for "disc jockey") has come to encompass many job titles, genres, and aesthetics.

"Everyone is a DJ," Jam Master Jay imparted. "When you're the one choosing music in the car, or loading up a CD player at a party; you're a DJ."

Certainly, everyone is at least an amateur DJ, in the same way that kids who sing in the school chorus or band could be considered amateur musicians. As I was growing up, my friends and I would make mix tapes for each other; mix CDs and iPod play lists could be considered the current expressions of the amateur DJ.

In addition, now there are software applications that let you combine, rearrange and remix your mix tapes in ways we could only dream of a few years ago. DJ hardware has advanced by leaps and bounds as well, and there is now a multi-billion dollar industry in place to support the desires of self-proclaimed amateur DJs to get closer to the music and make it their own.

Evolution is a tricky thing, since it never stops. Trying to put a finger on the evolution of the DJ is a bit like changing a tire while the car is still moving.

That said; let's take a look at where the DJ came from, how things evolved, where we may be headed, and what tools and skills are required to be a DJ.

PART 1

The Revolution

The Revolution of Records

First off, let's get our terminology straight. "Record" is short for "recording."

It has nothing to do with whether that recording is stored in vinyl grooves or digital bits. The Recording Academy still gives the Grammy for the Record of the Year, recording artists still make records, and DJs still play records. Because the vinyl record was the only game in town for a few years, many people came to associate the term "record" with its vinyl incarnation. However, a "record" is any recording made by a recording artist for release to the public. Wax cylinders were records. CDs are records. MP3s are records. DVD-As (Digital Video Disc, or Digital Versatile Disc-Audio) are records.

The birth of the record, like any good story, is full of conflict and unexpected turns.

The record's many incarnations and constant fight for survival is worth knowing, as there are many parallels to present-day issues. It started back in 1857 when French researcher Leon Scott de Martinville invented the "phonoautograph," a device that would record air pressure fluctuations by means of a diaphragm and a pig's hair bristle that traced a wavy line on a manually rotated cylinder. Unfortunately, this first recording had one big drawback: it could not be played back. It would take another 20 years until this problem could be solved.

Wax Cylinders

The big breakthrough arrived in 1877 when Thomas Edison, while experimenting with a new telegraph device, accidentally ran indented tin foil under a stylus. This led him to develop the first instrument that could both record and reproduce sound. Edison's machine used wax cylinders and was far from high fidelity. His invention was initially used as an office machine for businessmen, enabling them to record and play back messages and dictation.

Shortly after patenting his "phonograph" in 1878, Edison temporarily abandoned this project to concentrate on developing the incandescent light bulb. But by inventing the phonograph, he set off the initial spark, which would inspire further experimentation and eventually lead to the "wheels of steel" and beyond (Figure 1.1).

Fig. 1.1. The Edison wax cylinder recorder. This device collected sound through a horn, then translated sound pressure levels into movements of a diaphragm, which vibrated the connected stylus, cutting tiny grooves in a rotating wax cylinder.

And this is where Alexander Graham Bell enters the picture. After inventing the telephone (a device that would carry sound waves from one location to another) in 1876, it was a small step for Bell to develop what he called the "graphophone." With the money he was earning from his telephone invention, Bell established an electro-acoustic research facility in Washington, DC called the "Volta Laboratory Association," where he and others worked on improving Edison's phonograph. Their development used wax cylinders and a battery-driven motor, versus Edison's manually operated original version.

Flat Discs and Gramophones

In 1888, while Edison and Bell were embroiled in some ugly lawsuits over stolen patents, German immigrant Emile Berliner tried improving the existing models by using a flat disc instead of a cylinder. His master platter was made of zinc covered with a thin layer of acid-resistant wax that was scratched off during recording. He was then able to mass-produce records in vulcanized rubber. Berliner named his invention the "gramophone." These seven-inch discs had about a two-minute capacity and were manually turned by the listener at as close to 30 revolutions per minute (rpm) as they were able to get.

In the early 1890s, both the cylinder machines and the new disc-based gramophones were available and competing with each other, similar to the VHS verses Beta wars at the beginning of the video revolution.

Fig. 1.2. Disc-shaving machine. The shaving machine would smooth out the shallow grooves of the previous recording, making the wax cylinder smooth again and ready for new material.

The Columbia Graphophone Company was making some profit by leasing the first cylinder-based "juke boxes" to various entertainment facilities. The graphophone was no longer only a business instrument but had started claiming its place in the entertainment industry, with over 60 musical records available in the growing jukebox catalog. By the turn of the century, there were three major companies competing with each other: The Columbia Graphophone Company, Edison's National Phonograph Company (the two cylinder companies), and Berliner and Johnson's Victor Talking Machine Company.

One advantage of cylinder machines was their ability to both play back and record. The Edison cylinder machines came with both a recording stylus and a playback stylus, and various shapes of horns were available to contour the sound during recording and playback. The wax cylinders themselves could be recorded over up to a hundred times, provided they were erased properly using a disc-shaving machine (Figure 1.2).

Despite their better sound quality, cylinders were impossible to mass-produce and more cumbersome to store than discs. Berliner further improved the gramophone disc by utilizing an organic lacquer called "shellac," made from crushed-up beetles, and by creating the master entirely from wax instead of the zinc-and-wax combination. These two improvements gave the disc a much clearer and more dynamic sound. But sound quality was still miles away from being pristine. The discs could only reproduce the approximate frequency range of the human voice, and were unable to reproduce the more extended high- and low-end frequencies of strings or bass instruments.

The Victor label turned this shortcoming into a commercial success and found its market niche by recording famous classical vocalists. Being able to listen to these world-famous stars in your own living room became "one of the greatest pleasures of modern life," according to the ads.

The first hit record ever mass-produced was Italian tenor Enrico Caruso's 1903 10-inch disc on the Victor Red Seal label, the first record featuring the famous logo with the "His Master's Voice" dog.

Fig. 1.3. Kid Koala with his vintage Victrola. Notice the massive tone arm, the manual crank, and the storage space for records in the lid.

Victrolas, Albums, and Deccas

The cylinder's popularity took a big hit when the Victrola, a stylish and modern household record player, was released in 1906. It quickly established itself as the "must-have" home music product. Still, it took decades to put cylinders out of business for good (Figure 1.3).

Discs were now available in sizes of 7, 10, 12, 14, 16, and 21 inches, as well as double-sided formats. Tchaikovsky's *Nutcracker Suite*, released in 1909 on four double-sided discs, was the first record package called an "album," as it resembled a photo album.

Just when science, art, and modern life seemed unstoppable in their progress, World War I shook the planet in 1914. Instead of spelling trouble for the fledgling record industry, tough times put music in even higher demand. The first portable player, called a "Decca," enabled people to listen to music anywhere—even soldiers in the trenches.

It wasn't war that threatened to kill the record industry; it was the radio. If in the 1980s, "Video Killed the Radio Star," then radio killed the shellac star for a while in the 1920s. In the early 1920s, the Radio Corporation of America (RCA) started offering news, music, and entertainment for free through commercial radio stations. With the invention of the vacuum tube amplifier, radio reception and sound quality were improved. Live music broadcast over the radio sounded better and was more exciting to listeners than shellac discs on a Decca or Victrola, and record sales began dropping. The record industry needed a new point of attraction in order to win their audience back.

Technology to the Rescue

The early decades of recording consisted of sounds being collected into a simple horn, then mechanically etched into a disc or platter without the benefit of electronics. The condenser microphone developed by the Bell Laboratories in 1925 was capable of capturing the previously missing high- and low-end frequencies. Newly developed record players, known as "orthophonic sound boxes," came on the scene that, for the first time, included volume knobs, amplifiers, and loudspeakers. After some miss-starts, lawsuits, and clumsy beginnings, the record industry and radio companies put their rivalry behind them (see Chapter 2), and radio stations began playing the newer, high-quality records, and therefore, promoting their sales. Eventually, radios and record players were combined into home entertainment systems, which shared a common amplifier and loudspeaker.

Just when everybody was recovering from World War I, the great economic crash in 1929 took its toll on the modern world, and electronic devices became luxury items that only few people could afford. Small companies went out of business while new mass market (and less artistically-oriented) companies emerged out of the financial disaster. In the USA, the "American Record Company" (ARC) was formed, while "Electrical and Musical Industries" (EMI) took the lead in Europe.

Overlooking Stereo

The 1930s brought crucial technical improvements to the music industry, but it would take a while for this technology to make its impact felt. In 1931, EMI researcher Alan Blumlein invented binaural or stereophonic recording. A specially constructed motion-picture camera captured Blumlein's stereo recording of a steam train puffing by in 1935. One can see (and hear) this historic recording at Dolby labs in San Francisco, and it still sounds pretty great.

Amazingly, the recording industry didn't see the commercial potential of stereo, and EMI would let the patent expire before stereo swept the world 30 years later.

In 1939, magnetic tape was invented, but the tape recorder was still in need of some major tweaks. At the same time, jukeboxes produced by Wurlitzer now featured multiple selections and were taking over America's diners.

But at the height of all these innovations, World War II came crashing through the door. Would human hardship and financial disasters terminate the record industry this time? Quite the opposite. Radio transcripts and top hits on 12- and 16-inch records were sent to Europe and the UK from the USA in order to entertain and cheer up the troops. These special discs were named "V-discs" for Victory (or for Robert Vincent, who invented them) (Figure 1.4).

After the war, a tape recorder that far exceeded anything developed thus far in the USA was discovered in Germany, marking another milestone in the improvement of high-fidelity recording.

Meanwhile, shellac as the main material in the production of records was showing more and more negative aspects. It was brittle, easily scratched, and limited play time to 10 minutes per side. Additionally, it was impossible to increase the number of grooves on shellac without making the disc bigger than 16 inches. There were too many limitations, and it was time for a new medium.

Fig. 1.4. A 16-inch V-disc player and an early RCA microphone on display at a recent Audio Engineering Society convention.

Vinyl LP

Enter polyvinyl chloride, also known as PVC.

In New York on June 26, 1948, CBS announced the birth of the "Long Player," or the LP.

This new record was 12 inches wide, and turning at 33 1/3 rpm, it could hold up to 30 minutes of music per side. Because of its denser groove anatomy, it was said to be taking advantage of the "micro-groove technique." The following year, RCA introduced the first seven-inch 45-rpm micro-groove vinyl single disc.

By 1950, the 78-rpm shellac disc was entirely replaced by the new LP format. Although stereo recording techniques had been in existence since 1931, it wasn't until 1958 that stereo records were introduced, and many more years would pass before they became the standard. Consumer level quarter-inch tape recorders were capable of storing more information than vinyl, but record companies were so settled on vinyl as the main music sales medium that most decided to avoid releasing songs on quarter-inch tape, in order not to confuse buyers.

Fig. 1.5. A studer 24-tack multi-track tape recorder. This machine utilizes tape that is two inches wide and travels across the record heads at speeds up to 30 inches per second.

In 1961, EMI signed the Beatles, and the 1960s ushered in the golden age of the vinyl LP. Record sales skyrocketed. The demand for vinyl records was so high that manufacturers were having a hard time keeping up with orders.

Technological advances, including 4-, 8-, 16-, and eventually 24-track magnetic-tape recorders, improved microphones, recording consoles, plate reverb devices, and even synthesizers, helped producers such as George Martin, Phil Spector, and Phil Ramone bring record production to a whole new dimension.

Record covers became more complex and colorful, turning into collectible pieces of art. FM radio's improved fidelity proved a perfect match for album-oriented rock, and for about a decade, popular records fused youth culture, art, and politics to become the defining art form of the age.

Although a few manufacturers tried glove-box players, vinyl records and automobiles never mixed very well.

Eight-track tapes were short lived, but consumers who were anxious to take more control over their music and bring it with them on the road embraced the Phillips' compact cassette. The Sony Walkman, a portable cassette and headphone device, even made it possible to snub the world outside of your head.

LPs and cassettes ruled the 1970s and early 1980s, and many consumers became bedroom DJs, recording their own mix tapes for themselves and their friends.

The Compact Disc Digitizes Consumers

When CDs arrived on the scene in the early 1980s, first in Japan and later in the rest of the world, vinyl LPs began a steady decline, presided over by a record industry anxious to sell everyone their entire record collection all over again in this new digital format. Meanwhile, in the USA, MTV, VH1, and CMTV made video airplay more important than radio to new artist's careers. Prophetically, the first music video to be shown on MTV when it went on the air on August 1, 1981, was the Buggles' "Video Killed the Radio Star."

For a while, it seemed as though the major record companies had a license to print money. Video airplay, stadium tours, and the consolidation of radio markets helped create mega-stars like Michael Jackson, Garth Brooks, Madonna, Janet Jackson, Puff Daddy, Shania Twain, *NSYNC, and Britney Spears. Sales of CDs by just a few mega-stars became a huge portion of record labels' total sales, along with the re-issuing of old catalog on CD. This made moderately successful recording artists less important to the bottom line, and not something major labels felt compelled to spend a great deal of time on.

For DJs, CDs offered advantages. They were smaller and lighter than vinyl LPs, and not as prone to skips and pops and other physical wear. While most mobile DJs playing weddings and parties made the switch to CDs, the majority of Hip-hop and dance club DJs stuck to vinyl, which was easier to manipulate until CD turntables were introduced.

CD players became more widespread in cars, in boom boxes, and as portable Walkman-style players. Recordable cassettes lost steam. CDs became less expensive to manufacture, but consumers remained willing to pay a high price for them. Through most of the 1990s, CD burners were still way too expensive for average consumers.

But what technology gives, it also takes away. It could be that the easy money of the 1990s blinded the recording industry to the coming seismic paradigm shift. Why plan to change when things are going so well?

Digital Downloading and MP3

The conversion of music into digital bits ushered in the lucrative era of the CD. It also converted music into data, which put music squarely into the world of computers—a world where every 18 months, the speed of processing tends to double and the cost of data storage tends to half. In hindsight, the perfect storm created by the widespread adoption of the personal computer, CD/Rs, the MP3 audio format and the Internet should have been easy to see coming.

MP3 is a compression scheme, which allows good quality (though not quite CD quality) audio to be stored in drastically smaller digital files than CD's straight PCM (Pulse Code Modulation) system.

For music fans, this is great; you can encode your entire CD collection and carry it with you on a small MP3 player. The downside for the established record companies has been that anyone with a personal computer and an Internet account can download free (though illegal) MP3 files of practically any popular record ever made without even leaving their chair. For a few cents, they can burn scores of these music files onto a blank CD/R with the CD burner that came as standard equipment in their computer. Years ago, sales of blank CD/Rs surpassed sales of pre-recorded CDs.

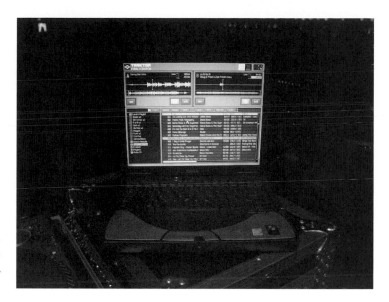

Fig. 1.6. The original Jazzy Jay's laptop computer, running *TRAKTOR*, puts a massive record collection at his fingertips, and interfaces with his turntables.

Just as we have seen earlier in this chapter, the new technology (in this case, compressed data files like MP3s flying over the Internet) is rendering earlier formats quaint, and the established business community (right on cue) missed the boat, and is scrambling to catch up.

Overcoming Cyberphobia

It took Apple Computer to design a business model that begins to tap the power of the Internet as a commercial distribution medium. For consumers, the prospect of any song being only one mouse-click away at any time of the day or night is a tempting proposition, making Apple's iTunes site an instant success. The runaway popularity of the sexy but expensive iPod, Apple's digital music player, once again drove home the point that consumers are happy to pay for something they really want.

The marriage of computers and music has other benefits. DJs can store thousands of records on the hard drive of a laptop computer, using one of the many handy DJ programs to organize, play, mix, scratch, and otherwise manipulate these records (Figure 1.6).

Musicians, producers, recording engineers, and re-mixers have an explosion of relatively inexpensive digital tools to aid in the creative process. Independent musicians have a new set of on-line tools, like MySpace.com, YouTube.com, Beatport.com, CD Baby, and iTunes, to help them build an audience and distribute their music. Underground artists making intelligent dance music (IDM), Hip-hop, house, jungle, and trance, along with various forms of indie rock, lo-fi, freak folk and punk, have been selling more records directly to their audience than ever before.

DVD and Beyond

Since the beginning of the new millennium, the excitement surrounding DVD (the fastest launching format in the history of home entertainment) and new super-charged video-game systems has made the plain ol' audio CD seem downright old fashioned, and not very sexy, in terms of bang for the buck.

DVD is a format designed to be adapted for many uses. Movies and music videos can be presented with excellent video quality and 5.1 surround-sound audio. While stereo employs two speakers (left and right), 5.1 surround uses five speakers (left, center, right, left surround, and right surround) and a dedicated sub-woofer (the .1), immersing the listener in a 360 degree soundscape, and potentially rattling their bones with bass. Interactive menus allow consumers to check out all sorts of extra features, which can include games, photos, alternate audio tracks and camera angles, bonus music tracks, and behind-the-scenes "making of" documentaries.

DVD-A and SACD (Super Audio CD) are formats intended to deliver stunning high-resolution audio quality to consumers with the proper playback equipment. BluRay and HD-DVD are higher storage capacity formats that dramatically increase video and audio quality even further, and increase the options that producers can. Like the Phonograph verses Gramophone wars of the 1890s and the Beta verses VHS wars of the early days of video, the propensity of the marketplace to introduce competing formats tends to confuse consumers when new technology is launched, and suppresses mainstream adoption. One could argue that the instant success of the LP, CD, and DVD formats was due at least in part to their widespread embrace by the vast majority of manufacturers and content providers, from the get-go.

In the early years of the new millennium, the record industry, pre-occupied with the decline of CD sales and the challenges of digital distribution, has been slow to exploit the new possibilities presented by the DVD formats. As we saw with the 30-year time lag between the introduction of stereo and the eventual embrace of the stereo LP as the industry standard, it sometimes takes a while for the potential of a new canvas to be realized.

For DJs, DVDs have helped make it possible to add video to their bag of tricks. The introduction of the DVD turntable, the Pioneer DVJ, gave DJs a way to harness the power of this new format, adding video to their sets. DJ-oriented DVD decks and mixers by Numark and others have also come to market, and video DJing, or VJing, has become a growth industry. Computer VJ programs like NuVJ, Modul8, and AVmixer Pro have also appeared, and like computer DJ programs, they offer VJs more storage capacity (via hard drives), built-in effects, and added flexibility and convenience.

Getting Better

The Internet, DVD, and other rapidly emerging technologies are ushering in a new era in which recording artists, DJs, and fans are deepening their relationships. The companies that figure out how to add value and help broker these relationships are coming along as well.

While the inevitable decline of CD sales has been the source of much doom and gloom during the first decade of the new millennium, there are bright spots in the music industry. Live concert revenues are up. Dance clubs across the world are filled week after week. Ring tones have grown to be a multi-billion dollar industry. DJ gear and related products continue to innovate and gain market share in the musical instrument market.

With mainstream music being perceived as more corporate and formulaic, the underground scene has been flourishing. With mainstream distribution channels currently undergoing a total metamorphosis, perhaps the most interesting chapter in the history of records is about to be written.

If history is any guide, the future of records is going to be incredible.

2

The Rise of the Radio DJ

Who was the First Radio DJ?

The first time a record was broadcast over radio waves was at 9:00 PM on Christmas Eve, 1906, from Brant Rock, Massachusetts, about 10 miles north of Plymouth Rock. Reginald Fessenden, a Canadian engineer who had worked for Edison and Westinghouse, managed to bewilder an audience of startled telegraph operators on United Fruit Company ships scattered about in the Atlantic Ocean.

Having no program director (PD) to dictate content, he sang (badly), played "Oh Holy Night" on the violin, read the Christmas story from the New Testament, and spun a wax cylinder recording of a slow piece ("Largo") by George Frederick Handel. Today, on a typical day in Brant Rock, 13 AM and 18 FM stations crowd the airwaves.

The first to eagerly claim the title of DJ was inventor Lee DeForest, who in 1907 played a record of the *William Tell Overture* from his lab in the Parker Building in New York. DeForest, whose invention of the triode helped make broadcasting possible, admitted later in life that there weren't many receivers in those days, but he would proudly assert, "I was the first disc jockey!"

The first female DJ was a ham radio operator named Sybil True, who hit the sparsely populated airwaves in 1914 with a show she called *The Little Ham Program*. She played "young people's records," which she borrowed from a local record shop, in order to turn youth on to the potential of radio. Her experimental show for young radio operators was prophetic, in that sales of the records she played on the air were constantly buoyed the very next day, to the delight of the store's owner.

Born of Controversy

Once commercial radio took off in 1922, it would take the record industry, publishers, and the union decades to allow radio DJs to legally play records on the air. Rather than embrace radio as a promotional vehicle, record companies placed warnings on record labels, forbidding their broadcast. The musician's union struck for almost a year to protest the practice of playing records on the radio.

The American Society of Composers and Publishers (ASCAP) forbade radio from playing any songs written or published by their members. In response, the National Association of Broadcasters (NAB) formed Broadcast Music International (BMI), which signed up many young country ("hillbilly"), folk, and African-American songwriters and artists to write and record records that could be played on the radio. This move had a significant impact on the development of popular music on the airwaves, giving an early boost to musical styles and musicians who may have been kept off the radio entirely, if radio's commercial potential had been known.

The first radio DJ to become a star and prove the massive power of a "disc show" was Martin Block.

A staff announcer at WNEW in New York, Block was covering the "trial of the century" in 1934—the kidnapping of the Lindberg baby.

Finding himself with nothing to do, during a long break in the proceedings, Block ran to the Liberty Record Store around the corner from the courthouse, bought an armload of Clyde McCoy records, and pretended to talk with McCoy as he stealthfully changed the discs on the air. He called the show *Make Believe Ballroom*.

When the advertising department at WNEW refused to lower themselves to find a sponsor for the show, Block urged overweight women who were listening to send in a dollar to purchase "Retardo Slimming Pills." The next day, 600 envelopes showed up, each containing a dollar. By week's end, $3750 had poured into the station's mailroom, and the new radio format had wings.

Martin Block came to epitomize the powerful radio DJ. Not only did he sell products for the show's advertisers, but if Block played a record on the radio, it became an instant hit. As he told *Billboard* magazine in 1942, when he played a record on the air, "If the platter is a good one, the most effective type of direct marketing has just taken place."

Black Music and DJs Take to the Airwaves

Also in 1942, *Billboard* launched a chart called the "Harlem Hit Parade." The chart later became known as "Race Records," until Jerry Wexler's term, "Rhythm and Blues," was adopted in 1949 to describe the rich field of popular music made by African Americans.

R&B gave rise to black DJs, whose numbers went from 16 nationwide in 1947, to over 500 in 1955.

Al Benson (AKA the Midnight Gambler), Eddie O'Jay, Hal Jackson, and Douglas "Jocko" Henderson (AKA the Ace from Outer Space) were among the most popular black DJs in the country.

Broadcasting from Harlem's Palm Café, Jocko Henderson, AKA the "Ace from Outer Space," would rhyme his way through his "Rocket Rhythm Review Show" with plenty of jive.

In their excellent book, *Last Night a DJ Saved My Life*, Bill Brewster and Frank Broughton offer this analysis:

Jocko, and similar loons, showed that the radio DJ could be a creative artist in his own right, not just a comedian or a companion but a vocalist, a poet. This aspect of the DJ's craft was to have momentous impact. In Jamaica, the sound system DJs emulated this jive rhyming almost immediately and became superstar deejays as "toasters" or "MCs." In New York 20 years later, there emerged the rapper, the descendant of both traditions.

White DJs also imitated their jive-talking black counterparts. Alan Freed took things a step further, and played black records on WJW radio in Cleveland.

The DJ Invents Rock 'n' Roll

Alan Freed promoted his "Moondog Coronation Ball" on the air, but worried privately that he wouldn't be able to attract enough paying customers to cover his costs. When an estimated 25,000 people showed up at the Cleveland Arena (capacity 10,000) to dance to the R&B acts that Freed played on his radio show, the police and fire departments turned on the lights and stopped the show, and the local press campaigned for Freed to leave town.

Many in Cleveland now point to this night—March 21, 1952—as the birth of rock 'n' roll. It was certainly a night that demonstrated the massive power of the radio DJ.

Freed landed in New York City in 1954, where he changed the name of his show to the "Rock 'n' Roll Party," and kept championing black music to a mixed audience. Freed used the terms "rock 'n' roll" and "rhythm and blues" interchangeably, and in a larger sociological sense, gave many white Americans their first exposure to Black culture. Once Elvis appeared on the scene, the term "rock 'n' roll" took on a whiter completion, becoming a euphemism for white folk playing black music.

The establishment didn't take well to excitable white teenagers being exposed to black music that seemed to be mainly about sex. Alan Freed was investigated by the FBI and congressional committees, and was eventually convicted of taking "payola," a common practice of receiving everything from cash to gifts to bogus writing credits in exchange for playing certain records. The link between record sales and airplay by popular DJs was now well established, and when Freed played a new record, it could easily sell 10,000 copies the next day.

While most other influential DJs were also guilty of taking payola, it's thought that Freed was hounded as much for his promotion of "subversive" music as for his shady business dealings. FBI Director J. Edgar Hoover ranted that rock 'n' roll was a "corrupting influence on American youth," and the congressional hearings would often fall into grandstanding by congressmen about the morality of this threatening music, rather than issues of bribery.

Rock radio DJs started losing the right to choose their own music when radio stations adopted a new position to their staff: the program director or PD. As much a market researcher for the advertising department as someone involved in musical aesthetics, PDs spelled the beginning of the end of radio as a spontaneously creative musical endeavor.

Playlists were cut drastically when research showed that the majority of listeners preferred to hear a few songs over and over again. The term "Top 40" was coined to describe this new format, and the predictability of a poll-based format quickly became a success with the advertising departments.

Risk-taking became the first casualty as PDs based music programming decisions on research telling them what people already liked, rather than trusting cunning local DJs to have a gut instinct about how their particular audience would respond to a new artist or style.

Album-Oriented Rock

There was a period in the late 1960s and early 1970s when some DJs threw out the restrictive formats and played album cuts from musicians who were pushing the boundaries. FM radio, with its higher fidelity and stereo capability, had at first been the domain of highbrow programming. As FM radio receivers started showing up in cars and clock radios, DJs with a more intimate personal style began playing music that took advantage of FM's superior sound and lower noise floor.

Tom Donahue, who owned station KMPX in San Francisco, pioneered this practice.

Donahue avoided the hits and cookie-cutter time restrictions in order to play bands from the underground Haight-Ashbury scene, like Jefferson Airplane and the Grateful Dead. Artists such as Yes, Emerson, Lake and Palmer, Led Zeppelin, and Pink Floyd filled the FM airwaves, and music with a high "cool factor" became referred to as "FM."

Eventually, this format dubbed "Album-Oriented Rock," (AOR) and joined the other formats with its own charts and demographic figures, which seemed by many to be contrary to its idealistic hippie roots. Marketing departments' demographic surveys sliced the radio dial into formats aimed at various segments of the population, maximizing the efficiency of advertising dollars. Along with Top 40, R&B, and Album-Oriented Rock, station formats came to include Urban, Country, Oldies, Adult Contemporary, Classic Rock, and Soft Rock.

In 1998, the Telecommunications Act lifted restrictions on how many radio stations a single company could own, eventually placing a few PDs or "consultants" in charge of a huge portion of the national airwaves. With the exceptions of college radio, public radio, and a few local stations and shows on commercial radio, music on US radio airwaves represents painstaking market research and high-stakes business dealings.

Since the music itself is no longer spontaneous, many commercial stations have gone with the paradigm of "on-air personalities" rather than DJs, putting a ring leader and a couple of side-kicks on to laugh it up during drive time. This removes the idea of DJ as music selector almost entirely, and while Howard Stern and other "shock jock" pioneers were once breaking new ground, the imitators seem to be struggling to keep things fresh.

Breaking into Radio

College radio stations present the best way for newcomers to break in, and many colleges, universities, and even junior colleges have radio stations connected to their communications departments.

To be a radio jock, you must learn to operate a mixing board, coordinate playing commercials off of one format and music off another, take phone calls, and talk on-air with a personality that draws in listeners.

Many stations script almost everything out for their radio jocks, from what to play when, to basically what to say. A few radio shifts are still around for creative individuals who pick and mix their own music, but the vast majority of career radio jocks go through many formats and locations.

Most radio jocks start out in smaller markets after college, and move around quite a bit for their first few years as they build up their résumés and gain experience. The pay for shifts in smaller markets is not high, but many local radio DJs enjoy the "big fish in a small pond" celebrity status, which can accompany the gig. "Drive time" shifts (radio's prime time, during morning and evening rush hours) on powerful large-market stations can pay a good salary. Job security is not high, as stations don't usually hesitate to shake up their roster and even their formats if the bottom line is not looking good.

The most successful radio jocks are those who have syndicated their own shows, and sold these shows to many different stations in multiple markets.

Future Radio Productions

Satellite and Internet radio are the newest entries into this narrative. Like HAM, AM, and FM, they enter the fray offering interesting musical alternatives, and are already embroiled in controversy and much legal maneuvering. As broadcast radio programming has become more analyzed, nationalized, standardized, and (some would say) sterilized, people are turning to satellite radio DJs, Internet DJs, and the independent live DJ to deliver them from the musical McDonald's that the airwaves have become.

Stay tuned...

3

The Rise of the Club/ Rave DJ

The dance club and rave phenomenon is a worldwide, multi-billion dollar industry that is often maligned and certainly misunderstood. Artistically, it does best slightly askew from the mainstream, morphing freely into subgenre after subgenre. Every decade or so, it is launched into the mass-market spotlight, where it's often hijacked by interlopers, adding to the confusion.

The story of the club DJ is also the story of the nightclub and the audience. All three coexist in an enigmatic, ever-changing triangle. Alter one side and something happens to the other two (Figure 3.1).

Fig. 3.1. The DJ, the club, and the audience form a symbiotic relationship.

So who is the audience—the club-goer, the dancer—to the club DJ? Who is the DJ to the audience? What happens in the spaces where they come together—literally and figuratively?

And why does a story about dancing contain not one, but two riots?

The Audience

Some are ravers—high schoolers, college students. Some are what society calls young professionals, in their 20s and 30s, with 9-to-5 jobs. Some are young, but not so professional—aspiring artists, actors, athletes, musicians, DJs, or writers, making ends meet at a day job somewhere so they can pursue their passion in between, on the side.

What brings them together is they like to go out at night, move to good music, have a good time in a room, club, warehouse, or field with a bunch of other people having a good time.

Maybe they go to the club to get lost (or found) in the music, to forget their other life. Maybe they go to celebrate: a new job, new love, new friends, the weekend, coming of age. Some say they go to capture that sense of community, of shared ecstasy.

Point is: they go. They dance. They listen. They get lost for a while on the dance floor. They take part in an ancient, time-honored ritual—dance—for all the reasons humanity has ever danced.

It's tribal. It's cultural. It's communal. It's spiritual. It's about transcendence, transformation, transfixion. It's about release.

To some, it's about *PLUR*: peace, love, unity, and respect. To some, it's about passion, play, sex, mating, and fun. It's about them—the audience and their self-expression through dance. It always has been.

You Must Be Jooking

The very first club DJs weren't DJs at all; they were members of the audience. Jukeboxes represent the first instances of people leaving their homes to pay money to hear records spun in a public place. In 1889, an Edison Class M electric phonograph in an attractive oak cabinet was fitted with a nickel-in-the-slot mechanism and four listening tubes. People stuffed 20,000 nickels into the device in its first six-month residency at a saloon in San Francisco. That works out to over 111 plays per day, seven days a week.

Most of the records produced in the 1890s were made to feed coin-operated players in public places. But jukebox sales surged into the stratosphere at the conclusion of two separate chapters in US history: prohibition and World War II.

A *jook house* was an out-of-the-way shack where southern field workers would go for dancing and drinking. The term "jook" was sometimes a euphemism for brothel. When a Texas distributor of Wurlitzer automatic phonographs began using the term *jukebox* around 1937, it took Wurlitzer headquarters a little while to clarify its meaning. Farney Wurlitzer banned the term, once he found out (Figure 3.2).

Before making jukeboxes, Swedish born Justice P. Seeburg made "orchestrions," which were automatic pianos with several instruments inside, designed to sound like an entire live band. In 1949, Seeburg's jukebox company came up with a mechanism that could play both sides of 50 records—an advancement so stunning that his competitors never caught back up (Figure 3.3).

Fig. 3.2. Jukeboxes became self-contained disco systems, combining flashing colored lights, state-of-the-art sound systems (at least, better than most patrons had heard before), and record selections constantly updated by regional sales reps or the owners of the establishments.

Fig. 3.3. Seeburg's revolutionary "Select-O-Matic 100" mechanism used two separate styluses to play both sides of 50 seven-inch records.

Those competitors, namely Wurlitzer and David C. Rockola (his real name), would limp along until the 1970s, when all three would be drummed out of business by human DJs, a rare instance of man replacing machine.

French Invent the Underground Discotheque

A fascinating chapter in the development of the underground dance club grew out of the popularity of African-American jazz musicians in Parisian nightclubs before the Nazi take-over in June 1940. The soldiers of the Third Reich immediately put a stop to jazz, seeing it as an unseemly collaboration between American Blacks and Jews.

The spirit of the thriving cabaret scene was not going to go away without a fight, so it was pushed underground.

Small, secret basement hangouts known as "discotheques" (French for "record libraries") became popular among members of the French resistance.

In some ways, these illegal nightspots resembled speakeasies in the USA during prohibition: memberships, passwords, and alternating locations. But live music was too big of a risk during the Nazi occupation, so discotheques played jazz records for the tiny dance floor—records so underground they had to be smuggled in by aficionados willing to risk being arrested or shot if caught.

When the war ended, discotheques went upscale. Places like the Whiskey au Go-Go and Chez Castel (complete with a VIP room) became precursors to the modern dance club.

White Kids Lose Control

Back in the USA of the 1950s, 45-rpm records known as *singles* or *45s* were starting to be spun at *sock hops*, which usually happened in the school gymnasium on the wooden floor of the basketball court. This location required the dancers to take off their shoes, which gave the event its name. If there wasn't a live band, radio DJs would often play the music, trying to smoothly introduce each song over the gymnasium's microphone, then moving that mic over to the speaker of the school's record player to amplify the sound. That sound was often boomy and scratchy, but most kids were too occupied with navigating through their raging hormones and remembering junior cotillion dance steps to notice, at least at first.

The dancing started to get a little more spirited as television's American Bandstand became increasingly influenced by black music, and the mainstream Dick Clark (arguably the first superstar VJ, or video jockey) followed Alan Freed in pushing for musical integration.

The sock hop was forever changed when the shameless teenagers of America started doing a radical new dance called "the twist." No more formal steps or polite behavior. No more dance partners or wallflowers. In short, no more rules. All you needed was the right record and a free spirit. The smash hit single (called simply, "The Twist") was actually a remake by an African-American teenager dubbed "Chubby Checker" by Dick Clark's wife, a play on the name "Fats Domino."

The twist set the dance floor free, unleashing a tsunami of new dances and making it possible for everyone, including the rhythmically challenged, to participate.

In London, Paris, and New York, new discotheques opened everywhere, catering to free-thinking dancers anxious to hit the liberated dance floor. In New York, a theatrical DJ named Terry Noel became a popular attraction at clubs like Arthur and the Peppermint Lounge. Noel outfitted the clubs with not one, but *two* turntables, in order to cut down on the space between songs, and is one of the first examples of a well-paid, respected, and celebrated club DJ.

Hippies Invent Disco

If you, like most people, thought that disco and the psychedelic, idealistic, West-Coast music scene that preceded it were diametrically opposed to each other, you were wrong.

The link is David Mancuso. As a child in Utica, New York, David spent time in an orphanage, where Sister Alicia would help the kids feel special and loved with mini-celebrations that included balloons, party food, and a stack of records.

Moving to New York City in 1962, in the midst of the Cuban missile crisis, David immediately took to the diversity and energy of his adopted home. He started to make some money in antiques. After a few years, he rented a big loft in a factory/warehouse district at 647 Broadway near Bleecker Street, and became a devoted audiophile, purchasing substantial Klipschorn loudspeakers, a McIntosh amp and pre-amp, and two AR (Acoustic Research) turntables. Since it was illegal to live in these lofts (Mancuso would hide his bed and cooking utensils from building inspectors), the neighborhood would clear out at night, making it the perfect place for a party.

The invitations for David's parties reflected his childhood memories, including a picture of small children sitting around a table, and the parties reflected his idealistic, hippie outlook that "music is love." There was always good food and a punch bowl, and the place was filled with balloons. In about 1970, due to an economic downturn, Mancuso started asking guests to chip in to cover costs, and the nucleus of the disco nightclub was born as a series of late-night rent parties at the Loft, which eventually became world famous (Figure 3.4).

Fig. 3.4. The fabled rent parties at the Loft were immortalized by Penelope Grill in paintings that capture the magic and diversity of the events.

Alex Rosner, a renowned audio designer who would later shape the sound systems for many important dance clubs, was friend with David, and helped in his ever-evolving exploits into high-end audio.

Their concept of making recorded music sound just as good or better than live music, by using the best sound system possible, is part of the Loft's legacy.

They also influenced Louis Bozak's designs for the world's first commercially available DJ mixers, the Bozak CMA-10-2D and CMA-10-2DL, now collector's items and beloved pieces of DJ history. The Bozak mixers used rotary pots, and components were chosen carefully to create the best audio output possible. The original Bozak mixers had no crossfader.

David Mancuso also founded and served as president of the first incarnation of the innovative New York Record Pool, which increased the power of DJs by giving them a voice with record companies.

Mancuso's belief in the power of playing a group of records for a tribe of dancers as a transcendent, healing journey was the genesis of the disco movement, and Mancuso himself is the archetype for the modern club DJ.

Mancuso didn't talk over records, like the radio DJs did. He didn't like to blend two records together unless the intro lent itself particularly well. He never used the pitch control, and was adamant about preserving the integrity of the original artist's intention. He didn't even consider himself a DJ—at least, not at first.

But present at the Loft parties, learning at his feet, were the saints and apostles of the modern dance club movement: David Morales, Frankie Knuckles, François Kevorkian, Tony Humphries, Danny Krivitt, Nicky Siano, and the legendary Larry Levan.

Larry Levan

A few months after Larry Levan passed away from complications of a weak heart and a decade of sometimes extreme substance abuse, *Vibe* magazine declared: "For over a decade, Larry Levan ruled the Dance Music world from his roost in the DJ booth at New York's legendary Paradise Garage."

Levan was, like Mancuso at the Loft, literally the Paradise Garage's resident DJ. For the first few years of the club's existence, he lived there. Converted from a parking garage, the Paradise was the pivot step between the Loft and the mega-dance clubs that would take over the world. Like Mancuso, Levan was more about vibe than chops; his technique was much criticized, especially in later years as substance abuse took its toll. But what every modern DJ learned from Levan (including a young Paul Oakenfold) was the importance of controlling a club, of manipulating the dance floor with music, of having an attitude.

It wasn't all about technique; it was also about creating an environment, a style, and a community with the crowd.

That community was diverse. Levan was of African descent, and the Paradise's wildly successful Saturday nights catered exclusively to a gay audience.

Before 1969, gay nightclubs in New York operated primarily underground. Many were unlicensed and often raided by police. But on Friday night, June 27, 1969, patrons fought back when police raided the Stonewall Inn, a popular gay bar in Greenwich Village. The violence and protests that followed became known as the Stonewall Riots, and marked the beginning of the gay liberation movement.

After Stonewall, many in the gay and lesbian population began to see dance clubs as a place to build community.

The New York dance club scene in the 1970s became a melting pot of hippie idealism intertwined with African-American, Hispanic, and gay cultures.

Francis Grasso

While Mancuso and Levin left the DJ, the clubber, and the club with the legacy of quality of sound and creating an environment, it was a DJ by the name of Francis Grasso that bequeathed to the modern DJ a bag of new mixing tricks.

Up to and including the 1960s and early 1970s, most DJs would play a series of short songs, totally unconnected to one another, often dropping the flow of the dance floor in the process. They were, in this way, not much more than a human jukebox.

Francis Grasso put the art into the art of DJing. He kept his eye on the dancers and the dance floor. Nobody mixed like him; he was a virtuoso daredevil.

Grasso perfected the art of live slip-cueing: holding a rotating record still while the turntable spins beneath, aided by a slip mat, so as to locate the best spot to drop in the new record, precisely on the beat.

He used a pair of turntables with speed controls, allowing him to match up records perfectly in tempo. Grasso experimented with equalization, liberally boosting the highs and the lows of a record to stunning effect.

He also used two copies of the same record to extend tunes. By blending two copies playing at the same time, he would achieve an echo effect, or by starting them slightly out of sync with each other, he created a phasing sound. He also superimposed two different versions of the same song (the studio version and the live version, or versions by different artists), one on top of the other with dramatic results. The more he gave out, the more enthusiastically the dancers gave back in a constant feedback loop of dancing and development.

Remixing

Many an early club DJ, out of necessity, had to invent and experiment. Take the case of the modern club DJ's bread and butter: remixing. A remix is a new version of the original song, most often built from the ground up by separating out the different strands of sounds from the original multi-track recording. Current DJs, such as Sasha, remix mega-hits for the likes of Madonna (*Ray of Light*) and Seal. The remix came about during the disco era when DJs transferred the live technique of extending songs onto tape and vinyl. Back then, pop tunes were short—a mere three minutes in length, designed mostly for radio play. The dance floor needed something lengthier, meatier.

Quite by accident, the 12-inch single was discovered when a mastering engineer was out of 7-inch acetates. Not only would the 12-inch format allow dance singles to be longer; the sound was superior, and the records themselves were easier for DJs to manipulate.

Larry Levin's first remix was a light-hearted disco version of "C is for Cookie" by *Sesame Street's* Cookie Monster. His versions of "Work That Body" by Taana Gardner and Instant Funk's "I Got My Mind Made Up" led to a string of remixes in the early 1980s that helped define post-disco dance records.

Post-disco?

If you, like many, knew disco mainly from the blockbuster success of the film *Saturday Night Fever* and the soundtrack by the Bee Gees, then much of what we've been discussing here—the dance club roots of true New York disco—may be new. There is no doubt that star power of John Travolta and the brothers Gibb did a lot to popularize disco.

But many in the scene felt that *Saturday Night Fever* had about as much to do with true disco as *Indiana Jones* had to do with archeology.

It may have also led to its demise.

Once *Saturday Night Fever* became a huge hit in 1977, dance studios across the nation were flooded with people wanting to learn to do the Hustle, and airport Hiltons, from Dallas to Dayton, hired DJs and hung mirror balls in their lounges.

Record companies promoted disco records to radio at a fever pitch, and moderately talented acts such as KC and the Sunshine Band ("Shake Your Bootie") and Rick Dees ("Disco Duck") grabbed the airwaves alongside more credible disco artists, such as Donna Summer and Chic. Rock stars who felt pressure to keep up with the times produced the most embarrassing (and overplayed) records of their careers, most notably "Silly Love Songs" by Paul McCartney, and "If You Think I'm Sexy" by Rod Stewart.

What happened next was the single most spectacular fall from grace of a musical genre from popular culture, ever.

Much has been written about the broad social implications of what came to be known as the "disco sucks" movement. While it's true that much of the urban disco movement consisted of "black female divas singing to gay men," chalking up mainstream America's rejection of disco purely to racism and homophobia implies that white males actually caught on to this fact. Perhaps some did, but they didn't get it from *Saturday Night Fever*, which was about a group of straight, 20-something, rather homophobic and racist Italian-Americans prancing about in ridiculous polyester suits.

The Bee Gee's falsetto warbling on "Staying Alive" (about being a lady's man) dominated radio, and overexposure started to feed a backlash against disco in general. When, with unprecedented hype, the Bee Gees moved in front of the camera to star in a pitifully weak movie based on the Beatles' near sacred *Sergeant Pepper's Lonely Hearts Club Band*, the US public was unforgiving. Travolta's next picture was a formulaic commercial for country two-step music and mechanical bull riding called *Urban Cowboy*, after which he too was banished for a time from pop culture, for egregious overexposure, along with the Bee Gees.

Disco anthems, designed for the New York dance club, didn't necessarily translate to the clock radio in a Missouri bedroom, or the car stereo in an Atlanta pickup truck. Once disco came to be associated with the mass media trying to cram the next trend down the throat of a gullible public, much of that public became mad as hell and decided not to take it any more.

Things came to a head in Chicago in 1979. White Sox promoters and Steve Dahl of radio station 98 FM (WLUP, *The Loop*) came up with the idea of a Disco Demolition rally at a double header between the Sox and the Detroit Tigers in Comiskey Park.

Those who brought disco records to destroy were admitted to the park for 98 cents, and the plan was to blow up a big box of disco records between games.

The flaws in the plan started showing up early as hundreds of records sailed onto the baseball field like Frisbees, many aimed at players.

Despite these bad omens, the pyrotechnics went ahead; fireworks were set off in front of a bin of records deep in center field, then a fireworks bomb was detonated in the bin itself. At first dozens, then hundreds, and finally thousands rushed the field. A full-fledged riot broke out as the crowd lost control and started to tear up the stadium. The police department's tactical force was called in and the second game had to be cancelled; the White Sox forfeited.

In their year-end issue, *Rolling Stone* magazine declared in 1979, "You can say that the first six months belonged to disco … and that the last six months belonged to the brave young rockers."

Disco records slowly stopped getting radio airplay, and the polyester suits went into the closet along with the Village People. True disco—namely, the dance club scene—just morphed and moved on.

Chicago House, Detroit Techno

Take Frankie Knuckles, the DJ at the Chicago club, the Warehouse, from 1977 to 1982. Frankie came from New York, where his compadres filled the void left by the fiery demise of disco with the new sounds of Hip-hop and electro. But Chicago was isolated, so when necessity came a-knocking, Frankie took to remaking songs on tape, before he went to the club.

He spliced and diced tracks to make them longer, more danceable, including intros, breaks, and vamps.

He fit weird songs or snatches of songs together, making a totally crazy, raw music that drove the dance floor mad. Little did he know he was creating the dance music Frankenstein called *house* (named for the laboratory in which he experimented) that still haunts us today.

Those that followed in his steps include Ron Hardy, house music's other mad genius, and the Hot Mix 5, a DJ collective founded by Farley Keith. Hardy preferred a sloppier, louder, edgier sound. Precisely because Chicago wasn't the cosmopolitan city of New York, house retained its dirty, funky edge, developing into a subculture and a style characterized by drum machines and synthesizers. Between 1981 and 1985, though, house came out of the shadows.

Cosmopolitan clubs from New York to, perhaps more importantly, London started to listen up and listen in, and house took off in the mainstream.

By 1985, the first house record labels were getting off the ground: Trax Records and DJ International in Chicago, and Easy Street Records in New York.

Shortly thereafter, Ron Hardy started spinning tracks by a Chicago electronic band called *Phuture*, which featured the squelchy sound of the Roland TB 303 bass synthesizer. Across Lake Michigan, the TB 303's cousin, the TR 909 drum machine, was getting a workout by Detroit producers working on their own musical style. Inspired by the Chicago house movement and influenced by European synth-pop artists like Kraftwerk, Detroit producers Juan Atkins, Derrick May, and Kevin Saunderson began churning out records that would become the blueprint for the techno genre.

The UK Catches Fire

In England, first Hip-hop and then house did a number on the club scene. House's fast-paced, progressive style quickly cannibalized the clubs. A generation of British DJs became remixers and producers, motivated by the new sounds of Hip-hop, house, and techno, and by the ever-decreasing cost of music technology (samplers, drum machines, and synthesizers). Energized by this new music, European DJs started making these genres their own, creating new, derivative sounds, including styles such as rave, ambient, epic, acid house, and hardcore.

The European dance scene absolutely exploded in the 1980s and 1990s.

The year 1988 became known as the second summer of love, after the 1967 blossoming of Hippie culture in San Francisco, when communal dancing (under the influence of mind altering substances to state-of-the-art sound and light shows) had swept youth culture and terrified parents.

In 1988, dances spilled out of clubs and warehouses and into the countryside around London, where huge open-air raves gave the DJ a heretofore-unimagined number of people to mix to and party with, in the strangest of places. The media ran sensational headlines about the dangers of new designer drugs (such as ecstasy), and police in the UK scrambled to crack down on the largely unlicensed events.

But the spark had caught a favorable breeze, and raves spread like brushfire.

France, Germany, Russia, Spain, the USA, then Africa and Asia all fell to the new incarnation of communal dance, where "PLUR" became the creed, bottled water the drink of choice, and DJs became revered as techno-shamans.

As the Bobbies worked to herd British youth back indoors, enterprising, young, soon-to-be-superstar DJs began creating their own career-defining club nights at famous clubs around the UK. Oakenfold started Spectrum Night at London's club Heaven, which spawned the equally revered clubs Future, Land of Oz, and Shoom. These clubs defined the acid house sound, which had adopted the Roland 303 as a staple element.

Renaissance in Mansfield was where Sasha got his start, and where he met his partner in crime, John Digweed. Together, they created a partnership that culminated in releases known as *Northern Exposure*. More than just a string of songs, *Northern Exposure* was the soundtrack to a night of clubbing, and the series has sold over one million copies worldwide.

Oakenfold went on to create the Balearic explosion from his regular club nights on the Spanish island of Ibiza. His contemporary, Carl Cox, whose résumé is equally impressive (the opening night of Shoom as well as numerous parties and illegal warehouse jams during the late 1980s' summers of love), took mixing to new heights with his dexterity on the decks. He was renowned for his ability to mix simultaneously on three decks rather than the standard two.

USA Raves and Festivals

In the USA, the popularity of raves boiled over in the mid- to late 1990s. The ideals of positivity and acceptance present at most outdoor raves were often in stark contrast to the elitist practices at many big-city nightclubs, where velvet ropes, guest lists, and VIP rooms separated people into classes.

The terms "rave" and "raver" began to fall out of favor in the early 2000s, probably from overuse. However, dance music festivals continued and became more mainstream events.

The annual Ultra Music Festival in Miami Beach, which happens in March during the Winter Music Conference, has become massive, boasting multiple stages and arenas, with different styles of music. The Ultra festival has featured DJ Tiësto, Paul van Dyk, Sasha and Digweed, DJ Dan, Fatboy Slim, Roger Sanchez, BT, and countless others.

Far from the mainstream, but a magnet for a large subculture that practices radical inclusion, self-expression and communal effort, is Burning Man. An eight-day alternative art festival and experimental community that springs to life on the playa of the Black Rock Desert in Nevada every September, the festival is named for the Saturday night burning of a wooden effigy. Art installations, odd temples, art cars, bicycles, and an instant city of over 30,000 people are all a part of Burning Man, which is about art during the day.

After dark, it's about the music. DJs keep the residents of Black Rock City dancing all night at provisional clubs like Root Society, Opulent Temple, and Deep End. DJs at Burning Man in 2006 included Freaq Nasty, Bass Necktar, and Scumfrog; Tiësto, Paul Van Dyke, and James Zabiela have attended in the past.

An entry from a book at the Burning Man Center Camp Café in 2006 read:

Burning Man Recipe

¼ cup Rave
1 cup Mad Max
2½ cup Circus
1 ton Alkaline
1 pinch Psychedelics
add alcohol to taste

Mix all ingredients in a Ziploc bag and cook for 119 hours at 105 degrees. Let cool for 2 days.
Nirvanna from S.F.

DJs Continue to Grow

Today, contemporary artist BT uses Macintosh Laptops and Musical Instrument Digital Interface (MIDI) controllers to whip crowds into a frenzy, controlling both video and music; mostly his own compositions. BT is also in demand as a film composer. Sasha wrote music for one of Sony Playstation's biggest games, "Wipeout 3." Oakenfold contributes music to movies like *Swordfish* and *The Matrix: Reloaded*. In fact, virtually all of today's DJs write and release original music.

The club DJs of today unrelentingly, unabashedly push the limits of music. Like their predecessors, they continue to experiment, bending sound according to the energy they feel reflecting back from the dance floor.

While many superstar musicians perform with the intention of impressing their audience, DJs seem to realize that the best musical events are those where the line between audience and artist becomes blurry.

That's when transcendence happens: when they both become not only each other, but also something else entirely. When the space between individuals shrinks, expands, even disappears. That's when the modern club DJ has tapped into the universal: all the people, parties, and endless experimentations that have come before.

3.1

BT's Adventures in Sonic Architecture

Calling BT a DJ is a bit like calling Woody Allen a clarinetist, or calling Thomas Jefferson a gardener. It's accurate, but only tells one small piece of the story.

BT (AKA Brian Transeau) is a musical renaissance man whose success encompasses many facets of the music industry.

BT's remixes and pioneering records in the progressive house and trance genres led many to assume he was a DJ long before he ever set foot in a DJ booth.

Some publications have even asserted that the term "trance music" was derived from his last name, an assertion he dismisses.

He's remixed tracks from high-profile artists including Madonna, Tori Amos, Sarah McLachlan, Seal, and electronic music pioneer Michael Oldfield. His producing chops gave boy band *NSYNC the hippest sound they ever had.

BT composed innovative orchestral film scores for *Under Suspicion*, *The Fast and the Furious*, and *Monster*, and wrote string arrangements for Peter Gabriel's *Millennium Dome* concert. Further associations with Peter Gabriel followed; Sting and DJ Sasha are among BT's other collaborators.

Brian began playing the piano at the age of four years, and studied arranging for strings as early as seven. He studied jazz at the Berklee College of Music in Boston, and is an accomplished guitarist and keyboardist.

BT travels the globe; both with live musicians and playing solo sets that blur the lines between DJ and live electronic musician, coaxing a groove-based symphony out of his laptops.

He brings a trained musician's ear and deep technology chops to everything he does, making his shows into a melting pot of computer music performance, live composition, and dynamic, cutting-edge groove evolution.

Always pushing the boundaries, BT's 2006 release, "This Binary Universe", is a collaboration with cutting-edge filmmakers and CGI artists, composed and mixed in 5.1 surround and released on Digital Video Disc, or Digital Versatile Disc (DVD) as well as CD. BT has added interactive video into his arsenal of live artistic expressions as well.

BT is also creating his own software company, Sonik Architects, with two new programming applications, Break Tweaker

Fig. 3.1.1. BT tweaks a parameter on an Oxygen8 Keyboard during a live performance in southern California.

and Stutter Edit, prototyped by BT in cSound and Supercollider, slated to be the first releases.

BT has been spending time at Berklee, visiting classes and presenting workshops. What follows are my conversations with BT, as well as some excerpts from his presentations.

What was your first computer system?

My very first system was a PC Model 1 with 16K of RAM and a green monochrome monitor. I used to write in COBOL and BASIC, which is where I got my love for programming.

Did you do any music on this system?

I did, actually. The first sequencing program I ever got, and what I did my first album on, was Voyetra Sequencer Plus Gold. There are still some things I miss about that program, believe it or not. It's a fantastic program.

It seems like every time I go to a new program, there'll be stuff that totally blows me away, and there'll be stuff it doesn't do.

But the timing, as far as Musical Instrument Digital Interface (MIDI) goes, on the PC and later on my Lunchbox 386 PS/2 Model 70, was actually really, really good. The MIDI fired almost directly off the CPU; there wasn't an interface. Also, the PCs were the first computers to use crystal sync clocks. They're in all the Macintoshes now, but they weren't in the Mac Classics and stuff.

Going back and looking at the waveforms of my earlier recordings, the timing is exceptionally tight, as far as MIDI goes, and it was all stuff that was done on those earlier PCs.

A lot of English people talk about the Atari computers like that. I have a lot of friends who are English and still use Ataris. Those computers literally fire MIDI data directly off the motherboard. There's no separate interface; there's MIDI sockets on the side of an Atari. And the timing, as far as MIDI goes, is great.

Let's segue into an issue I know you have some feelings about, and that's "MIDI spew."

Yeah, that's a big one for me. I don't know if you've looked at how sloppy MIDI is, but this is a protocol that was invented in 1981, and these days, a USB mouse communicates with the computer faster than MIDI. It's really messy, when you try to send multiple controllers down a 3-MHz cable.

My whole issue with MIDI timing is how it disproportionately skews data.

"MIDI latency" is a total misnomer. It's not just latency. If MIDI feels like firing an event 50 or 200 samples early, it'll do that, and some events will happen late.

I remember a test we did a couple of years ago, with a Korg Trinity. We were just firing 16th notes, using a metronome sound, at 120 beats per minute, on a single channel. When you look at where the notes fall, the timing is bad news, man. And that's a modern MIDI synthesizer that's used on loads of records. It's a great-sounding synth, too.

I started noticing the timing discrepancies when I bought my first Macintosh and got my first Sound Tools system, back in 1995. That's when I stopped using MIDI. On my first ProTools rig, I started recording individual audio tracks, and that's when I developed my time-correcting techniques.

I'm sure some may say, "Hey, what's the big deal? A 100 samples here, 50 samples there—who's going to notice?" What's the advantage to having "sample-accurate" grooves?

The problem with MIDI spew, for the sort of music I do—for dance music and film music—is that the listening environment is large. In a club environment, the space itself exponentially expands upon however bad the latency or slop is. Obviously, it's most dramatically noticeable with your drums, bass lines, and percussive things, which have hard attacks.

If you're listening on a pair of NS10s or a pair of speakers at your house, you don't hear time discrepancies that much, and you don't hear flamming and things of that nature. But when you play tracks in a large space, if the alignment of the kick drum and bass line isn't right, then the attacks are mushy.

When you time-correct a whole track, it just sounds phenomenal. It's a cumulative effect, and you really hear it.

It's a very complicated process to time-correct audio and make it feel and sound natural. I've made up all these techniques for doing groove templates, dealing with pre-attacks, all sorts of stuff.

A lot of DJs aspire to become musicians. You're a musician that became a DJ.

That's true, it's totally backwards! It's funny, because calling me a DJ—although I love doing it, it's really such a misnomer. But I was called a DJ for so long ....

I remember interviews where you were insisting, "I'm not a DJ!"

As a joke, my manager made T-shirts for me last Christmas that say, "I'm still not a DJ!"
Everyone always thought I was a DJ, and I used to contest it in interviews in a funny way.

All my friends are just exceptional DJs, and I thought it was insulting to them to call me a DJ.

I could appreciate it on an artistic level, but I liked performing electronic music.

What made you finally break down and decide to pursue becoming a DJ in your own right?

When I really broke down was when some technological advances started being made that I couldn't ignore any longer. It made me realize that there's going to be a median point coming very soon between live electronic performance and DJing. I wanted very much to be a part of that.

So, what I do is kind of this symbiosis between electronic music and DJing. I will play other people's tracks, and my own tracks, but I'm remixing and writing tracks on top of them live.

It's a really exciting thing to do, and I like it a lot because it requires such focused attention.

Playing with a band, I can have a conversation while playing guitar or piano or synths, but this is like, "No, I really can't talk, I really need to concentrate!" So, it's challenging, and I like that, too.

When did you start playing out as a DJ?

Around 2001. I could beat-mix for a long time, but kind of just playing around. I have two turntables and a mixer at my house, but I never sat around and played records. My friends are always sending me the most amazing records months and months before they ever come out, so I started pressing them, made a set, and went out and started doing it.

This is with vinyl records?

Yeah. I went out and did it traditional-style, with vinyl, to really get my feet wet. Once I started getting the hang of that, I gradually migrated off vinyl and onto the laptops. I still play vinyl, especially if I haven't had time to digitize a file. Also, it's fun beat-mixing vinyl records. The tactile sensation of dragging your finger on the record or pushing it … it's a good feeling.

Now I'm doing gigs primarily off of two TiBooks and some Oxygen8s. I'm running two programs, and it's actually going to grow to three because I'm going to be DJing video at the same time. I've been digitizing all these QuickTime clips, using this program called Arkaos, and working with an R-Chaos developer on implementing some really musical things with it. It's basically like an engine for synthesizers and patches, and within a patch is a QWERTY keyboard. Each key fires a different QuickTime sequence with different effects, and the effects are controlled by MIDI information.

How did you arrive at using multiple laptops, and what programs are you running?

I found that it was pretty CPU intensive to run Live and Reason, or Live and Reaktor, on a single computer. And it's scary; you don't want the TiBooks to choke during your show.

With two, you have a backup if anything happens.

Yeah, exactly. It's nice for screen shots, too, because you can see what's going on while not needing to flip between programs. Each TiBook is running one program, and they're linked by MTC and a couple QUATTROs.

What do you have the Oxy8s set up to do?

I use them for triggering loops. So, for example, I've made sample-accurate, bar-long loops that I've imported into Live. What I've done is set up 16th-, 32nd-, and 64th-snare fills running through a filter with a crash to cut them off, so I can do

Fig. 3.1.2. Apple Macintosh Laptop computers power BT's live sets.

impromptu fills in a track, and high-pass filter that track while it's playing to make a track that didn't exist in the record.

I've got all kinds of live percussion loops, kick loops, and break beats. I do a lot of tonally related mixing because you can pitch-shift the tracks in Live but keep them the same tempo.

For the gig last night, I warp-marked "Day Tripper" by the Beatles and "Trans Europe Express" by Kraftwerk. I was teasing the "Day Tripper" track through a resonant low-pass filter and through a delay. People went crazy with that. With Trans Europe Express on top of it, it was dope!

So, it affords you the opportunity to do a lot of hyper non-conventional things, working off the laptop, 'cause then, you're not locked into key and BPM. Anything's accessible to you.

Do you enjoy traveling as a DJ?

That's the coolest thing. DJing around, you get to travel to all these really exciting places. I've been to Ibiza [an island off of Spain] many times. It's so much fun. It's a place though, that you can only go for like two days, because it's so insane. I've never seen anything like it.

How many nightclubs would you say there are on Ibiza?

Big ones, like the Space and the Posh, there's about 10. But there's a lot of cool bistros and stuff. If you had it in America, you'd be like "this is the coolest place I've been to in my whole life."

This club, Space—it's crazy.

When the sun rises the roof is electronically rolled back and planes fly overhead, and it's louder than the sound system, it's deafening. People cheer the planes coming in, "Welcome to Ibiza!"

It's an insane place.

Another club, Pasha, has a double-Olympic-size swimming pool in the middle of it. I've stage dove into the pool during my sets. It's crazy, man!

The sound systems and the clubs there are amazing, and you can do really long sets. I would never play this long, but I've been to watch someone play for like seven or eight hours, before. Some DJs will do 12- to 15-hour sets.

Could you contrast playing in Ibiza to playing in Tulsa last night?

It's a whole different thing, but equally as rewarding. I played in a room last night where there were probably around a thousand people, and it's in Tulsa, kind of a cool small-town vibe.

You can go to a place like Ibiza, or England, and play at a place with 10-, 20-, 30,000 people, and it'd be great, energetically. But there's something to be said for playing for even 500 people, where it's meaningful to everyone—yourself included. It means as much, if not more, than for 20,000 people to be into it and then go "Oh, let's watch the next thing".

Fig. 3.1.3. Many of BT's live sets mix traditional DJ techniques with elements of live electronic music performance.

How will you play the two gigs differently in terms of how you construct your set?

That's a good question.

When I'm playing for a really large crowd, I'm much more conscious about doing something that has a thematic through-line.

If I'm playing for a big crowd, I'll play some sort of techy house stuff, and then gradually build into progressive stuff over the course of two or three hours.

But if I'm in a place where people are expecting me to play a lot of my music, it's actually a lot more fun. So, I'll play some new-school, break-beat stuff, some break-step stuff. I'll even play some drum 'n' bass. I'll play some trance, some progressive stuff, and some deep house. There's more freedom to take it in different directions.

Let's say you get to the club and the DJ before you is killing the place, and you take over. What do you do?

That's a good question, too. Playing live, doing electronic music with a bunch of sequencers and drum machines and all the rest of it, could be a real problem, if you have a DJ on before you playing gabba.

If they're literally playing hardcore at 200 bpm, and you come on playing at 128, you'll be like, "Dude, this sounds like crap!"

I often like to put on a record as someone is finishing their set. I like stopping, too. Delineating. "Give it up for this guy! He did a great job."

Even if I do beat-mix into his song, I'll keep it pace-wise where he's at, but I'll always bring it down after that. So if he's playing progressive stuff, I'll play something that's around 135–136, with more harmonic stuff going on in it. I'll gradually bring things down with more minimal tracks.

Then, maybe five records in, I'm at my starting point, but without telemarketing that to the crowd.

Much of the DJ aesthetic revolves around remixing, which is a big part of what you do. A goal of many DJs is to become a remixer, which you were first.

That's very true. Remixes were the first thing that I ever did that came out in a public form. Before my first record came out, I did a remix of "Not Over Yet" for a project named *Grace*.

How did you approach it?

I did that at a friend's studio. He had a Macintosh, maybe a Mac IIFX. It was running Logic Audio with a NUBUS version of ProTools. I guess that would have been Logic 3, around that time.

I brought a couple of my keyboards over; I had a D70 at the time. Victor, whose studio it was, had a couple Juno 106s, a 909, an 808, and a Sequential Circuits Drum Tracks. I love the claps on those.

So, I remixed "Not Over Yet." At the time I did it, it was very unique, for what it was, and it got a great reaction.

How was it unique?

A lot of dance music up until the early 1990s was very linear. What I brought to house music, which was progressive house and ended up being called trance, is this kind of non-linear aesthetic, very influenced by classical music and film scoring, harmonically, but more so dynamically.

I remember playing that track for a couple of my friends—the Deep Dish guys, Adley and Sherom. There wasn't a typical breakdown, but there was a passage that was 64-bars long, which was one of these big, huge, ambient things with big swirling phasing pads. I introduced the vocal in that part and pedal-toned up until that part, and then introduced the chord progression. There was a big buildup into this huge, crashing chorus, where the beats come back in.

Adley and Sherom listened to it, and they said, "What the hell are people gonna do on a dance floor during all this ambient music? You have to keep the beats going."

And I said, "I don't know; I like the way it sounds!"

I'll never forget my first trip to England, and DJ Sasha played that record as his last song of the night.

To see people with their hands in the air, screaming, during what was, for all intents and purposes, a piece of ambient music, was really a revelatory experience for me. It was like, "Wow, people want to hear dynamics in this kind of music."

And all of the early things that I did, right up to now, are hyper-dynamic.

You and Sasha have gotten to be good friends.

He's incredible. I've never seen someone who intuitively is able to mix things in the same key or related keys the way he does.

Say you had a record that's at 128 bpm, and is in D minor, and then you have a record that's 136 bpm and is in E-flat. If you meet in the middle, there will be a midpoint where both of the things will be in the same key.

He is able to do stuff like that just completely by ear. I've heard him mix for hours in the same key or related keys, and I don't even think he's conscious when he's doing it.

Fig. 3.1.4. BT visits a Music Synthesis class at his alma mater, Berklee College of Music in Boston. What follows are excerpts from BT's master class later that day.

Dance-club music is generally pretty tame, harmonically, although you've been able to stretch things a bit. What are the harmonic parameters, and why do they exist?

I try to incorporate interesting harmonic sensibilities to stuff, and it's hard to do.

The thing is, especially with house music, pedal tones sound great, because there is a frequency relationship between the kick drum and the bass line. But a bass change following a harmonic progression can dynamically alter the relationship between the kick drum and the bass line.

Depending on what? The specific playback system's subwoofers, and the resonant frequencies of the club?

Exactly. And it can end up sounding, for lack of better words, "cheesy." So, trying to work within that framework, and trying to do things that sound interesting harmonically, excites me a lot. I love it.

On film scoring:

When I got the job to score *The Fast and the Furious*, it was one of the first films where I got to incorporate some of my more traditional knowledge, combining viola, cello, contrabass, brass, and woodwinds with some of the more esoteric electronic skills that I've used doing dance music for the last 10 or 11 years. This was one of the most open gigs I've had.

The director, Rob Cohen, is a very musical guy. I can talk about bar articulations, and he's savvy enough to understand to some extent what I'm talking about.

For those of you who have seen the film, it's a bunch of hyper-stylized, MTV-cut, crazy, fast-and-furious scenes. The name makes sense.

I wanted to use things that subliminally reminded the viewer of motion and movement and machinery, so I said to Rob, "I want to notate traditional percussion parts, piatti, orchestral bass drum, and cymbals, and transpose them to car parts."

I was expecting to get laughed at, but Rob was really cool with me and said, "I think that's a great idea, I'll deliver two car chassis to you."

So, he sent them over.

We had a cool studio out in the valley that had a really big room, with big doors that we could open to wheel in these massive cars' chassis.

Instead of using brush cymbals, we hung up hubcaps and used snare brushes on hubcaps. Instead of using orchestral bass drum, we'd drop a trunk door.

The theme is a hybrid of strings—violin, viola, cello, contrabass—and some brass. It features one of my favorite brass instruments, the chimbasso, which sounds like evil drum 'n' bass lines when you have five or six of them playing together.

Fig. 3.1.5. BT, a pioneer of sample-accurate time correction, demonstrates specific techniques during his master class.

We did this theme in several sections. First, I recorded the orchestral sections, then I recorded the car-part percussion, and then I did electronic treatments over top of it.

On time correction

Time-correcting tracks to the sample-accurate level really make for powerful sounding tracks, when they're played in a large space. I've spent a lot of time coming up with a way to do sample-accurate time correction of audio. It's a simple technique, but it involves a lot of personal preferences. It's a complicated process to make it feel and sound natural.

This technique is irrelevant unless you shelve [EQ] things properly.

You can have the punchiest attack transients in the universe, and have one keyboard pad that's got sub-harmonic information in it, and your entire mix will sound muddy. Start shelving frequencies from 150 Hz and lower out of everything other than your kick drum and bass line.

It really makes a tremendous difference in the overall sound of the composition—especially in club music.

I converted a lot of my friends who work in MIDI to sample-accurate time correction. Sasha, for example, who works in dance music, worked in MIDI for years and years and years. I started showing him some of these techniques, and he's like, "Okay … it's in time, but so?"

He played a track that we worked on, before it was time corrected, and he'd say, "Oh, it sounds so sloppy on a big system."

Then I sent him my time-corrected version of it, and he was like, "Can you come back and show me how to do that?"

Now, the Hybrid guys time-correct all their stuff. Sasha and the guys that work with Sasha time-correct quite a bit of stuff.

One of the biggest things, stylistically, is that this type of time-correcting is about losing a lot of pre-attacks.

A lot of acoustically played material has pre-transients or pre-attacks as part of the sound.

Compression sounds great. We all love compression, and it makes drums really thump. But it adds really bad attack transients that sound sloppy if you stack a lot of loops and/or bass lines. If we zoom into this waveform, there will be this beautiful kick drum, and there will be this ... "thing" in front of it. And what is that thing? Take that away, we don't want that there! When you take it away, it sounds a lot better.

Things like acoustic guitars don't sound natural if you cut where the guitar is actually ringing. So, to balance using some attack transients is a very difficult thing.

You can imagine how long this takes. Some of the songs from my new album took about a month of Mike (BT's assistant) and I time-correcting away.

Some of the songs have a lot of live playing in them. I did some songs with

Fig. 3.1.6. Individually treating minute slivers of audio in radically different ways gives BT's tracks a signature sound.

Tommy Stemson (the Replacements), and Richard Forest (Guns and Roses), and Brain (Primus). We cut a bunch of tracks together, and those guys are incredible players.

Part of the reason why this started for me was trying to merge acoustic and electronic instruments, and intertwine something and make it sound musically useful and groove but not be sloppy. So, we spent literally a month on some of the performances, to try to get them right.

On going virtual

I just recently moved my studio. I have a lot of really beautiful analog, vintage equipment that I've collected over the years, and I've put it in storage. I'm going to try and see if I can live without them, and see if I can function completely, or almost completely, in the digital domain. Some of the Virtual Studio Technology (VST) instruments that are coming out are absolutely astonishing.

I just found a plug-in called Exciton and another one called Plastic from a company called reFX. They are two of the most exciting sounding synthesizers I've heard in my life.

The plug-ins are crazy, I want to hold them! It's weird; it doesn't seem real, but it sounds so good.

On stutter editing

I spent a lot of time developing a complex process I call "stutter editing," which basically consists of repeating certain pieces of audio and then treating the individual pieces as you go.

Sometimes, I'll have a bar or two bars in a composition where I will literally spend eight to ten hours on a fill or turnaround. It satisfies no one but me, but it is a lot of fun to do.

I'll spend a lot of time, especially with vocals, doing really interesting treatments, but trying to aid the composition. Take things and pitch them around. If you drop something by an octave, it's going to be half time. It might sound sludgy, or it might sound really interesting. You might be able to use pieces of it, or you might be able to use a quarter note of it, but it could sound really, really cool.

For the track that I un-apologetically did for *NSYNC, I did over 40 treatments to their vocal. I used Kyma (by Symbolic Sounds) and I ran Justin beat-boxing through a Marshall amplifier. And then I sat for, literally, like a week, trying to make Justin sound like a cross between Michael Jackson and Max Headroom.

It doesn't really matter unless it's moving to you. And all you can hope for is that if it is moving to you, that it is moving to other people, too.

There's no point in playing something or doing something just for the sake of technical proficiency. You can show off all you want; you can shred on an instrument, or you can do the most amazing stutter edits in the world. If it doesn't aid a good composition, in the end, it doesn't mean anything.

So, I encourage you to explore this with reverence for your compositions. And reverence for what you are trying to achieve emotionally.

3.2

Paul Oakenfold

Paul Oakenfold has been called the most successful DJ in the world. He's won "Best DJ" titles from practically every magazine that awards such titles, earns more money, sells more records, and jet-sets around the world to play cooler clubs, theaters, and stadiums than practically any other DJ in history. He's remixed everyone from Madonna and Tori Amos to Snoop Dog, scored movies like *Swordfish*, and contributed music to *The Matrix* series of films. He helped popularize trance, acid house, Goa, and the Balearic spirit of the Spanish island of Ibiza. He and his production partner, Steve Osborne, have even received a "Best Producer" nod at the Brit Awards, the UK's equivalent of the Grammys.

Fig. 3.2.1. Paul Oakenfold has been expanding what it means to be a DJ for well over a decade.

Oakenfold's career started in London in the late 1970s as an A&R man for UK-based Champion Records. He showed his penchant for success early on, when his first signing for the label was Jazzy Jeff and the Fresh Prince (AKA Will Smith), and his second signing was Salt-N-Pepa.

He is credited with changing European youth culture in the 1980s and early 1990s, through his pioneering DJ exploits, production work with the rock band Happy Mondays, and remixes of UK bands New Order, the Cure, and Massive Attack.

In 1991, Oakenfold's remix of U2's single "Even Better than the Real Thing" reached higher in the UK charts than the band's original version. He joined U2 on their historic ZOO TV tour, and remixed "It's a Beautiful Day," which reached number 1 on the US dance charts.

He was the headlining DJ on Moby's ambitious *Area: One* tour, and his record *Perfecto Presents Another World*, when released in 2000, became America's biggest-ever DJ mix album.

In 2002, Paul Oakenfold released *Bunkka*, his first album as a solo recording artist, which features guests as diverse as Ice Cube, Nelly Furtado, gonzo journalist Hunter S. Thompson, and Shifty Shellshock of the Los Angeles rock–rap band Crazy Town.

While no one with as much success as Oakenfold is without their detractors, his skill as a live DJ has been honed through decades of hard work, and his ability to rock a party is beyond question.

I spoke to him about live DJ skills, remixing, building a career, and his evolution from DJ to rock star.

What's the largest audience you've played for at this point?

A hundred-thousand, hundred and ten. In Italy.

Your tour with U2 back in the early 1990s; was that the first time a DJ toured with a major rock act like that?

It was the first time, yeah.

Did anyone think it was crazy and it wouldn't work?

Sure, all the old rock 'n' roll heads. They couldn't figure out what that was all about. But it was the band's idea, and it comes from them, really.

How did you meet up with U2?

We were hanging out in Dublin, and then they kind of said, "Look, would you be up for this?" And I said, "Yeah."

Very low key?

Yeah. I just remember the first show I did, it was pourin' rain, and I was out on stage, and my records fell off the stage, 20 feet down to the ground and got soaked. That was a nightmare in front of 35,000 people.

Oh, my God!

Yeah, Bret's CD got stuck, that was playing at the time. Oh, it was just a nightmare.

What did you do?

I just let the CD play, and then we tried to jump down and grab the records and threw them back up.

Ended up, after we finished, we went back into my dressing room, and we borrowed a blow dryer and blow-dried all the records.

That was my first experience of being on that tour.

You recommend that new DJs practice their technical skills and technique. What technique currently comes in most handy in your sets?

Probably extending sections of the record. Take two copies of an obvious record that you know, that everyone knows, and extend the break, the intro, the chorus, to make it different.

Do you accomplish this through back spinning?

No, I'm doing it in more of a structured way. I'll mix one record 16 bars beyond the other one. Bring it in, wash it in, wash it out, and stay 16 bars behind. So, it's extending 16 bars.

If you wanted to take it even further, play the first record for 16 bars, then mix in the second record, that would make it 32 bars, and you could take the needle off the first record and repeat it again. Just repeat the same section as long as you want.

Can you give me an example of a record that you'll use this particular technique with?

There's *Seventh Son*. It's my own record, so it's something that I'm really, really familiar with. I can just pick up the needle and drop it at any point and know exactly where it is.

If you're a practicing DJ, and you have the time to learn the structure and the arrangement of every record, you can do that. But I really don't have the time to do that on every new record, anymore. The only reason I can do it on my own records is 'cause I know exactly where it is.

Do you remember where you first picked this up?

It's been going on for years, really. Hip-hop DJs, I presume, started it. The old-school DJs were doing it, so it's nothing new. It's just a good technique that really works for the crowd.

What new skills have you been exploring recently?

I'm into more kinds of effects, right now.

I like to use a lot of filters on one record while I'm mixing in another. For example, I'll filter the bass out of the record the crowd can hear, and then bring in the bass of another record.

I'll put the needle on top of the other record, and just play the bass line. So basically, I'm changing the bass line of the record that's playing out there.

What gear do you favor these days?

I have a rider and a spec, so if I'm booked to play a club, they have to get the equipment I need. Which is a Rane mixer and three SL12 turntables with 6/8-inch diamond-cut studs.

Where do you recommend aspiring DJs begin?

Well, I would say, they should get out there and watch other DJs play, and practice at home.

What's the most unusual thing you do to practice?

I don't practice. [Laughs] I used to, but now I don't have time to. I do check records out, now. On the road, I'll go to sound check before the shows and see.

One of your tips on how to become a successful DJ is to start your own club, or your own club night. What makes a successful club night?

The crowds. I mean, that the crowd and the music are doing something different. I can obviously only comment on how I did it, and that was exactly what we did.

The suburban area we lived in, we started our own club. We were playing different music than everyone else, and it developed and went from there, really.

You've been associated with many different genres of dance and club music. In your eyes, how did all these different styles evolve into what they are now?

Development. Just a developing sound over a period of time.

The great thing about dance music is that you can be experimental, because it doesn't cost much money to produce realistically.

Compared to other genres of music, you can be your way and come up with these sounds. I mean, that's that drum 'n' bass game. Just an existing sound that kept getting faster and faster.

How would you describe drum 'n' bass to someone?

I would describe it as 160 bpm. I think it's the same as half the speed and double it.

How would you define acid house?

It's based around a [Roland] 303 sound that's quite annoying, actually. Well, it's not just a sound, it was more a way of life, as well.

Tempo-wise?

Tempo-wise, like 121.

What about progressive, how would you describe that?

More melodic sounding, with lush strings and pads.

Trance?

Trance is uplifting, riff-based, with a melodic, hypnotic sound.

What tempos would you relate to these genres?

House would be 120 to 127. Progressive, 130 to 135. Trance, 135 to 140. Techno is 140 and up.

Could you talk about the concept of bridge records?

Each bridge record is a key record, because that's where you can realistically change genres. You can go from progressive to break. And also, they are records that you can change the key. So yeah, they're important records.

Do you find yourself building around bridge records, thinking a few records ahead of time to take it somewhere else?

Yeah, sometimes. I do think a lot about how I'm going to into my break section and out of my break section. Those two records going in and out of those sections, and blending house and whatever you're playing before and after, are very important.

Tell me about your record *Bunkka*, and what you set out to accomplish while you were making it.

I wanted to teach myself by doing an album that was based on songs. A couple years ago, when I decided to make the record, the music out there wasn't inspiring me, and I wanted to do something that was different. That's why I decided to do a record that was based on a harder route to take—something that I felt would be more rewarding than doing an instrumental record.

What was the process of writing the material?

It's pretty straightforward. You come up with an idea, then we would go into the studio and put down a rhythm that would work, and then build from that rhythm. We might have samples, depending on the kind of direction we wanted the song to take.

I mean, I certainly don't have anything against instrumental records. As a DJ, I'm a big fan of instrumentals. I love them. I mainly play instruments.

But to do an artist record, I wanted to do something different. And that's the reason why I went down that road.

How did your collaboration with Hunter S. Thompson come about?

Me being in contact with him. I'm sure Hunter didn't know who I was. It was just through a mutual friend. And then I kind of explained what I wanted to do. Why I thought he was important to use. Then we went from there.

The single "Starry Eyed Surprise" got heavy airplay on MTV and radio. Can you tell me how your collaboration with rapper Shifty Shellshock came about?

I was playing in a nightclub in Seattle. They were hanging out and saw what was going on in the club, and we just decided to do something together. That was kind of how it started. We got together and came up with that tune. I had this rhythm, and I needed to find a lyric that would work. It was based on everyone having a good time. We were just kickin' back and having a good time.

Have you done many videos before this?

Not really. It's a difficult experience.

I'm not really a key fan of doing video. But, it's something you get thrust into and you have to take part in.

How did the tune you do with Nelly Furtado, "Hard As They Come," come about?

The inspiration came from when I was working on the *Swordfish* soundtrack. That was based on a very down-tempo kind of vibe, and we wanted to do something that was more cutting-edge and dark.

Tell us about remixing U2's "Even Better than the Real Thing."

The most important aspect of remixing, from my point of view, is all about keeping the integrity of the artist. That's really what it's all about.

So, I would work with as many live musicians as possible. For "Even Better than the Real Thing," I spoke to the band, and I told them exactly what I was going to do. I changed the complete rhythm and the structure, keeping the same key. It was a pretty straightforward, not complicated situation, in that they just let me get on with it.

After I changed the rhythm, which is programmed drums, I played the keyboards. And then got in a bass player who was better than me to play, and a girl to do backing vocals for chorus.

We came up with a keyboard line that was all around the riff. It was very important to have riffs, 'cause that's what U2 is really all about.

We kept as much of U2 that I thought would work and went from there.

Played it for the band; they liked it. They wanted a few changes on the guitar part. I said to them, "Of course, but they do kind of work." It was cool.

What format did you record to?

We used a 24-track through an SSL (Solid State Logic mixing console). We used it to do basically do all our mixes. I've also tried a Neve, which has got warmer sound.

We used to edit to tape. No more.

Are you using ProTools these days?

Yeah, every once in a while.

Do you like using ProTools?

No, I prefer the SSL, because I learned on an SSL. There's something about a studio with a big console that I kind of like, rather than just sitting in front of a computer, doing it anywhere.

I don't know …. It's not what it should be, for me personally, when you're in a little room, just being on a screen, rather than in a big studio.

How does your experience as a DJ affect your remixing?

The DJ side is built all around rhythm. When you're remixing, you're going to change the rhythm for dance music. You're directly in touch, so you know first-hand with what works and what doesn't.

With "Even Better than the Real Thing," we did a rough mix. I played it in the clubs, and the intro wasn't long enough. I like a 16-bar intro, but it needed 32.

I wouldn't have changed it unless I'd played it out and seen that it didn't quite work for the record. It needed to breathe before the vocals came in, and it didn't. The vocals came in too soon. So, I changed that, and it was purely down to being out there most weekends and seeing what does and what doesn't work.

4

The Rise of the Hip-hop DJ

DJs were Hip-hop's original architects, and remain crucial to its continued development. Hip-hop is more than a style of music; it's a culture. As with any culture, there are various artistic expressions of Hip-hop, the four principal expressions being:

- visual art (graffiti)
- dance (breaking, rocking, locking, and popping, collectively known in the media as "break dancing")
- literature (rap lyrics and slam poetry)
- music (DJing and turntablism).

Unlike the European Renaissance or the Ming Dynasty, Hip-hop is a culture that is very much alive and still evolving. Some argue that Hip-hop is the most influential cultural movement in history, pointing to the globalization of Hip-hop music, fashion, and other forms of expression.

Style has always been at the forefront of Hip-hop. Improvization is called *free styling*, whether in rap, turntablism, breaking, or graffiti writing. Since everyone is using essentially the same tools (spray paint for graffiti writers, microphones for rappers and beat boxers, their bodies for dancers, and two turntables with a mixer for DJs), it's the artists' personal styles that set them apart. It's no coincidence that two of the most authentic movies about the genesis of the movement are titled *Wild Style* and *Style Wars*.

There are also many styles of writing the word "Hip-hop." The mainstream media most often oscillates between "hip-hop" and "hip hop." The Hiphop Archive at Harvard writes "Hiphop" as one word, with a capital H, embracing KRS-ONE's line of reasoning that "Hiphop is a culture with its own foundation narrative, history, natives, and mission." After a great deal of input from many people in the Hip-hop community, I've decided to capitalize the word but keep the hyphen, to show both respect and deference to decades of tradition.

Background

To understand any culture, it's helpful to look at the political and economic factors that faced the civilization that created it.

In 1945, Robert Moses, an autocratic New York state municipal official, proposed the Cross Bronx Expressway. His plan would displace tens of thousands of lower-income families and necessitate the building of massive housing projects to replace existing tenements.

"I dare say that only a man like Mr. Moses would have the audacity to believe that one could push (the expressway) from one end of the Bronx to the other," expressway designer Ernest Clark in an interviewer from the PBS series *The American Experience*.

In the 1950s and early 1960s, scores of Bronx neighborhoods were leveled to build the eight-and-a-half mile stretch of highway, which to some became the symbol of "white flight" from the city to upscale suburbs in Westchester County and Connecticut. During Governor Rockefeller's administration, the imperious Moses was finally ousted from his state job, but the events he had set in motion were irreversible.

The Bronx became an area of destroyed communities, abandoned buildings, piles of rubble, and austere new high-rise housing projects surrounded by asphalt basketball courts and chain-link fences.

Tough economic times in the 1970s put New York City on a restrictive budget that had little money for social programs, and no money for music or art in inner-city schools.

Gangs clashed with police and each other. The job of survival made concepts like art and civic pride seem a luxury.

In the excellent small book, *The Rough Guide to Hip-hop*, Peter Shapiro contends:

The first generation of post-CBE (Cross Bronx Expressway) children in the Bronx was the first group to try to piece together bits from this urban scrap heap. Like carrion crows and hunter-gatherers, they picked through the debris and created their own sense of community and found vehicles for self-expression from cultural ready-mades, throwaways, and aerosol cans.

All they needed was a leader, and they found Hercules.

Kool DJ Herc

In 1955 in Jamaica, a young woman from the parish of Saint Mary gave birth to a son who would become the father of Hip-hop. As a child, Clive Campbell was inspired by local Jamaican DJs, who would set up their massive sound systems for outdoor parties in the open spaces (called "lawns") around Kingston, and enhance the dance by toasting on the mic over the instrumental sections of the records they played.

Top Jamaican DJs were fiercely competitive with each other. The size of their sound system and freshness of their records were especially important. DJs would name their sound systems. Duke Reid's was "the Trojan," while Prince Buster put together "the Supertown" sound system.

In order to play records no one else had, Jamaican DJs made record buying trips to the USA. Eventually, the major Jamaican DJs raced to become record producers. They recorded local talent to have exclusive, danceable tracks to feed their sound systems, which led directly to the development of ska, rock steady, reggae, and dub music.

While young Clive Campbell was especially fond of records by the Godfather of Soul, American R&B singer James Brown, Clive's Jamaican roots shaped his concept of the role of the DJ and how to rock a party.

Immigrating to the Bronx in 1967 when he was 12 years old, Clive was growing up tall. At Alfred E. Smith High School, he spent a lot of time in the weight room, which earned him the nickname of Hercules. Like a lot of his friends, Hercules was a graffiti writer, running with a crew called "the Ex-Vandals." They were also getting into rocking, but the DJs playing the parties were not catering to this explosive new energy on the dance floor the way he and his young friends thought that they should. Hercules knew he could do better.

One could argue that the first Hip-hop events happened in the community center of Clive's building, 1520 Sedgwick Avenue.

Taking on the DJ name of Kool Herc, Clive charged the guys 50 cents, and the ladies 25. Herc's impressive presence (and mutually respectful relationship with gang members) helped keep order, and his choice of music helped draw the crowds.

At first, Herc was playing through "PA columns and guitar amps," but he quickly plowed his earnings into building up a massive sound system capable of bone-rattling bass. The Jamaican DJs he remembered from his childhood all named their sound systems, and Herc dubbed his "the Herculords system."

Another Jamaican influence was his work on the microphone. Herc would give shout-outs to his friends in the audience and extol the virtues of his sound system, often in rhyme (*à la* Jamaican toasting style) over the instrumental portions of the records he was spinning.

Eventually, Herc turned his attention to increasingly complex manipulations on the turntables, and let his friends Coke La Rock and Clark Kent take over on the mic. They became known collectively as "Kool Herc and the Herculoids," and were probably Hip-hop's first DJ/MC crew.

One turntable technique Herc pioneered was the practice of extending "breaks." The break was the part of the record when everything dropped out except for the drums and percussion, and sometimes the bass. This section was usually only four to sixteen bars in length, but it made the best part of the record to rhyme over, and the B-boys and B-girls (the name bestowed upon the virtuoso dancers, also known as "breakers") would save their best, most crowd-pleasing moves for the break.

Herc got the B-boys off the sidelines and became their favorite DJ by playing just the breaks (or breakbeats) of the records, instead of the entire record. The "Clap your hands, stomp your feet" section of James Brown's "Give It Up or Turn It Loose," was one such favorite, as was the drum part on "Apache," by the Incredible Bongo Band.

Herc explained that he would go right to the "yolk" of the record, leaving off all anticipation and just playing the beats.

By many accounts, Herc was the first DJ to take the stage with two copies of the same record in order to repeat a particular break by switching between two identical records. While he didn't switch back and forth in perfect time, this concept of taking a portion, or "sample" of an existing record in order to create a new piece over which to rhyme (or rap) was the birth of sampling and looping, two of Hip-hop's core musical concepts. Another Hip-hop staple Herc established is heavy bass with a sparse track, definitely an influence of his Jamaican background.

By the summer of 1973, Kool Herc was setting up his sound system outside, like the Kingston DJs of his childhood.

Instead of lawns, Herc and other early Hip-hop DJs would hold block parties on the asphalt basketball courts and playgrounds adjacent to neighborhood schools. The B-boys would show off their moves while the Herculoids rocked the mic, and Kool DJ Herc threw down the grooves—not just the stiff, overproduced disco sounds favored by the white DJs in the palatial clubs in Manhattan, but obscure funky records you had to dig for. These were records so sure to drive the crowd wild that Herc soaked the labels off so that competing DJs wouldn't be able to find them. Records with *soul*.

The crowds were large and young. There were no movie theaters in the projects, cable TV and video games were still years away.

In many ways, the birth of Hip-hop was the beginning of a ghetto Renaissance.

The term "Renaissance man" depicts someone who is into many disciplines; its equivalent in Hip-hop culture is the "B-boy." Often used to describe a virtuoso dancer, the "B" can stand for "beat" or "Bronx," and often depicts someone skilled in multiple expressions of Hip-hop: graffiti, breaking, beat boxing (the act of emulating a drumbeat with your mouth), rapping, etc.

Afrika Bambaataa

Impressed by Kool Herc's block parties, young Afrika Bambaataa saw the potential of turning this burgeoning scene into a positive force with an international scope. He and his friends started to throw parties at the Bronx River Community Center, dragging their home stereos into DJ for the gatherings. Setting up on opposite sides of the room, Bambaataa and his friends would signal each other with flashlights as their records were ending, so they could keep the music going non-stop like DJ Kool Herc.

Bambaataa was a born organizer. He was also a member of the Black Spades, one of New York City's notorious street gangs. Inspired by the image of Africans fighting off colonialism in the 1963 movie *Zulu*, Bambaataa co-founded an organization called the Zulu Nation, and took the name of a 19th century Zulu King, Afrika Bambaataa, which means "Affectionate Chieftain." It was 1973.

Bambaataa threw himself into building the Zulu Nation, with a charter of promoting "freedom, justice, equality, knowledge, wisdom, and understanding."

The Zulu Nation organized dance and DJ competitions and musical events, promoting peace and racial tolerance. Hip-hop crews provided an alternative to street gangs, and DJ and dance competitions were less dangerous than outright gang banging, even though fights would sometimes break out.

While violence was still a gritty reality in the poverty-stricken ghettos of New York City, the prodigious era of graffiti artists' murals on New York trains actually coincided with a significant decline of hardcore gang activity.

DJ Afrika Bambaataa quickly became known as the "Master of Records," combining tracks from almost every genre into an eclectic mix full of surprises. "Bam" was one of the first of the young Bronx DJs to join the record pools, a system set up by the record labels to get new releases into the hands of influential DJs. This further expanded his growing collection, and turned him on to the quirky German techno band Kraftwerk.

Fig. 4.1. A recent photo of graffiti on a freight train. Graffiti like this is once again on the rise.

Once record labels started to take notice of this new underground scene, Afrika Bambaataa produced "Planet Rock" (1982), a hit single containing samples from Kraftwerk's record *Trans Europe Express*. "Planet Rock" launched a style known as electro funk, and was a big influence on Detroit techno and other genres of dance music.

Bambaataa has toured the world promoting Hip-hop as an international movement, encouraging the locals to rap in their own language about issues important to their own lives.

Life magazine featured Afrika Bambaataa in their "Most Important Americans of the 20th Century" issue, citing the positive work of the Zulu Nation, and Bam's pure original vision of Hip-hop culture: peace, unity, love, and having fun.

Grandmaster Flash

Joseph Sadler was another youngster who took inspiration from Kool Herc's block parties and breakbeat innovations.

Sadler also studied the work of Pete DJ Jones, a more polished disco DJ whose seamless beat mixing was indicative of the exploding Manhattan club scene. When Pete Jones let young Joseph sit in on his system, Sadler discovered the concept of cueing up the next record in the headphones, a technique that made beatmatching possible.

While headphone cueing was already a standard practice among club DJs, Sadler's mixer didn't have this capability. Sadler was an industrious student of electronics, so he designed his own cue system, built it with parts from Radio Shack, then Krazy Glued it to the top of his mixer, and called it his "peek-a-boo system."

Sadler set to work practicing, intent on developing a new level of DJ skills. Taking the name Grandmaster Flash, he combined the seamless flow of Pete Jones with Herc's practice of repeating just the breakbeats and climaxes. In the process, Flash developed several new techniques, and redefined what it meant to be a Hip-hop DJ.

Many point to Grandmaster Flash as being the first DJ to approach the turntable as a musical instrument.

Flash was the first to mark his records with a line to show the beginning of a cue, something he dubbed the "clock theory." This allowed him to seamlessly extend any musical section using two copies of a record, perfectly keeping the beat. Flash developed and named the "quick mix theory," and executed it by counting the exact number of backspins necessary to repeat specific numbers of bars or beats. Practicing with two copies of *Lowdown* by Boz Skaggs, Flash got to where he could reconstruct arrangements on the fly.

He put together a set of only "the best parts" of a string of records, but when he first unveiled his new techniques in front of an audience, the response was anything but enthusiastic. It would take months of gigs for Flash to hit his stride and find his audience, but once he did, he was celebrated by legions of fans, including Debra Harry (AKA Blondie) who wrote and recorded "Rapture" in tribute to Flash.

MCs also gravitated toward Flash, and after packing Harlem's Audubon Ballroom with close to 3000 fans in September 1976, Flash and his crew of MCs were at the vanguard of this new underground movement, which would eventually be known as Hip-hop.

Flash organized his MCs into *Grandmaster Flash and the Furious Five*, a crew of rappers consisting of Cowboy (Keith Wiggins), Kidd Creole (Danny Glover), Melle Mel (Melvin Glover), Mr. Ness (Eddie Morris), and Rahiem (Guy Williams). They developed many of the house-rocking practices, phrases, and rhymes that became standards of the Hip-hop party, along with the Cold Crush Brothers, featuring DJs Charlie Chase and Tony Tone, MCs Grandmaster Caz, JDL, Easy AD, and the Almighty Kay-Gee.

Early on, when fans and entrepreneurs would recommend to Flash that he and his crew make a record, Flash would wonder out loud why anyone would buy a record of some guys rhyming over other records. Rapping was something that happened at parties and nightclubs.

Everyone was taken by surprise in October 1979 when an unknown crew called the Sugar Hill Gang scored a hit with "Rapper's Delight" on Joe and Sylvia Robinson's Harlem-based Sugar Hill Records. The hit proved there was an audience for rap records, considered up until that point to be a live phenomenon.

Flash and the Five responded immediately with a record of their own titled *Superrappin'* on the fledgling Enjoy label. They moved to Sugar Hill Records in 1980 to reach a wider audience, and found that audience with the party rap record *Freedom*.

In 1981, an entirely different kind of record, aptly titled *The Adventures of Grandmaster Flash on the Wheels of Steel* showcased the culmination of turntable artistry up until that point in history.

Performed live by Flash alone at the decks, the seven-minute collage still stands as a masterpiece of cutting, mixing, and in-your-face creativity.

Flash and the Five also released *The Message*, considered by many critics to be rap's first socially challenging record, as well as *Flash to the Beat* and *White Lines (Don't Do It)*, before splitting up.

Fig. 4.2. Grandmaster Flash at a recent gig.

Grandmaster Flash is playing better than ever as of this writing, and if you get the chance to see him live, don't miss the opportunity. The Old School doesn't get any better than Grandmaster Flash rocking a Hip-hop party.

Grandwizard Theodore

As a kid, Theodore Livingston developed "needle dropping," a technique that involves looping a section of a single record, in time, by lifting the tone arm and consistently dropping the needle an exact number of grooves earlier.

"My mother had a phonograph that she played in the living room," explains Theodore. "When she wasn't around, I used to take 45s, like 'Scorpio' by Dennis Coffey or James Brown records like 'Funky President,' and when the break part came, I used to skip the needle back and forth and forth and back, not knowing that my needle dropping skills were coming into play."

Theodore's brother, DJ Mean Jean, was Grandmaster Flash's DJ partner for a while, and Flash sometimes had young Theodore demonstrate his amazing needle drop technique during his sets.

Many credit Grandwizard Theodore as being the DJ who invented scratching.

Like Newton's apple, scratching found Theodore by accident.

"It was 1975, I was twelve and a half or thirteen, living with my mother on 168th street and Boston road. I was basically in the house trying to make a tape. Back then we didn't have no tape deck—you had to put the (boom) box in front of the speaker on top of a chair and hope that you got the clearest sound possible."

"My mother is the kind of person where, she doesn't play; she just comes out swinging. So I was playing two records, and she bangs on the door, and told me: 'Listen, either you turn the music down, or you turn the music off.' While she was in the doorway screaming at me, I was playing one record on my right-hand side, and I was holding the other record with

Fig. 4.3. Grandwizzard Theodore and the author after a show at the Berklee Performance Centre.

my left hand. I wanted to keep the groove going, so I was moving the record back and forth while the record on the right-hand side was playing. When she left, I realized what I was doing. Sometimes you find yourself doing something you've never done before, and you just keep practicing it and perfecting it."

While Flash and others had used similar techniques (which he referred to as cutting) while cueing the top of a track or a break, it was Theodore who practiced up on scratching as something to do over another record in a musical context. As Theodore further developed scratching and needle drop techniques, he would use them on gigs with his brothers, performing as "The L Brothers." Eventually, everyone gathered around the turntables to see what the heck Theodore was doing.

"They gave me the name Grandwizard because when I play a record, I would always play around with it," Theodore recalls.

Grandwizard Theodore came into his own with his MC crew, the Fantastic Five, playing successful shows at the T-Connection, Ecstasy Garage, and Harlem World. The Fantastic Five and the Cold Crush Brothers kept a rivalry going, which both groups played up. This rivalry led to some pivotal DJ/MC battles where the applause of the audience would determine which group would take home the cash prize of $500 or $1000. Theodore and the Five recorded *Can I Get A Soul Clap/Fresh Out of the Pack* in 1981.

Theodore was the first DJ to link up with the Red Bull Music Academy in 1998, and appeared in the 2001 film *Scratch*. He currently teaches at the Scratch Academy in New York, and plays gigs around the world. When he's in New York, he plays regularly at Willie's Entertainment and Club Mumbai.

Grandmixer DXT

Grandmixer D.ST introduced scratching and the whole idea of being a turntablist to the world.

Called "D" by his friends, he would go to shows where Kool Herc was DJing, and started spinning as early as 1974. He took the moniker of D.ST for D Street (Delancy Street), where he used to hang out. D.ST became a B-boy who could dance, rhyme, and spin.

After honing his DJ skills at house parties, D.ST became a premiere Hip-hop DJ, and was the first to spread that culture into Westchester County at the Mount Vernon Boy's Club. Back down in the Bronx, D.ST played the T-Connection and the PAL (Police Athletic League), before landing a gig DJing weekends at the prestigious Roxy Theatre and night-club in Manhattan.

"The Roxy became the place to be, and all the A-list New York City actors and entertainers went there to hang out. I was the guy spinning, and people started approaching me from all over."

The Roxy raised D.ST's profile considerably, and lead to gigs in Paris and London. D.ST also got the attention of a young bass player and budding producer named Bill Laswell, who was becoming interested in Hip-hop and scratching.

A drummer and multi-instrumentalist comfortable playing bass and keyboards, Grandmixer D.ST brought an especially musical approach to the turntables. When Laswell had the opportunity to produce a Herbie Hancock record, he brought in Grandmixer D.ST to play the turntable as if it was an instrument in the band.

The single "Rockit" featured D.ST's rock-solid, confident scratching as the main solo.

While "Rockit" didn't reached number 1 in the *Billboard* charts, the music video was one of MTV's first huge hits, and the video's bizarre, artsy visuals helped it sweep the first-ever MTV Video Music Awards. D.ST toured in Herbie's Rockit Band, playing the turntable (or as he dubbed it, the "turnfiddle") like a musical instrument, employing effects and breaking new ground.

The success of "Rockit," including its live performance on national television at the Grammy Awards, took the Hip-hop insurrection of scratching and launched its spores into the collective consciousness of youth culture. Thousands of these spores took root and eventually returned in the form of the Invisible Skratch Piklz, Beat Junkies, X-ecutioners, Fifth Platoon, and most scratch DJs who were between ages eight and twenty-eight in years 1984.

Fig. 4.4. Grandmixer DXT and the author at the historic Milestones Awards event in 2006.

Grandmixer D.ST renamed himself Grandmixer DXT in 1989 after the death of his Brother. He continues to tour and make records (some of them with Bill Laswell), and contributed original music and sound design to the soundtrack of the film "*The Untold Story of Emmett Till.*"

Branching out into other areas of production, Grandmixer DXT founded Transfer Master, a company that restores and re-masters analog master tapes through a process DXT named "Forensic Editing." He recently restored the highly acclaimed "Thelonious Monk Quartet with John Coltrane at Carnegie Hall" record, distributed by Blue Note.

Jam Master Jay

Jason Mizell began playing drums and singing in the choir at Universal Baptist Church in Brooklyn when he was five years old. In 1975, his parents moved the family to Hollis, half an hour east of Manhattan, in the southeast corner of Queens.

While those living in the Bronx projects considered Hollis to be the cushy suburbs, Jason still had his hands full at Junior High School 192 and Andrew Jackson High, where he once dragged his buddy Wendal Fite to safety after Wendal had taken a bullet in the leg.

Still, Hollis was a step up with its small single-family dwellings, many with basements, aluminum siding, and postage-stamp sized front lawns. Future secretary of State Colin Powell's parents had also moved to Hollis to bring up their kids.

Hollis also produced its fair share of Hip-hoppas, including LL Cool J, Q-Tip, Run (Joseph Simmons), DMC (Darryl McDaniels), Russell Simmons (Run's older brother), and Ja Rule. Wendal Fite even grew up to be DJ Hurricane, spinning and scratching for the Beastie Boys. In Jason's own words:

I wanted to be a drummer, then I saw Grandmaster Flash. That changed everything. When Hip-hop came to the neighborhood, it was like basketball—you had to know how to do it.

Jason's mom bought him a Technics SL-10 turntable, and he would practice in his headphones late into the night.

The large stretch of asphalt next to Junior High School 192 known as Two-Fifths Park was where Jason (AKA Jazzy Jase) honed his skills in front of an audience.

Someone would force open the base of a streetlamp and hot-wire an extension chord to power the equipment, and eventually, as many as 500 people would turn out to party.

Jazzy Jase vanquished many DJs who dared to battle him, and he developed a keen set of scratching, cutting, and beat-extending skills, including a cross-arm technique, which became one of his crowd-pleasing signature moves. This led to Jason landing a weekly gig at Dorian's, where on one night in January 1983, over a thousand people showed up.

Jason was miffed at two people who hadn't made it out; his MC friends Run and DMC had spent the night recording "It's Like That" and "Sucker MCs" in producer Larry Smith's attic studio. When Russell Simmons got them signed to the fledgling Profile Records as Run–DMC, they invited Jason to come on as a full member of the group. DMC came up with the name "Jam Master Jay" for Jason, and the first single they cut together as a trio was "Hard," with a track called "Jam Master Jay" on the flip side, where the two MCs rapped about the skills of their DJ in old-school Hip-hop style.

In 1984, Profile compiled the singles into the first-ever rap album, called simply *Run–DMC*. The video that accompanied the record, *Rock Box*, was the first rap video on MTV, and the album sold over 500,000 copies by the end of 1984. Their second album, *King of Rock*, became Hip-hop's first platinum album (over one million copies sold), eventually selling over four million copies.

Jam Master Jay and his rock-leaning record crates were also responsible for another major trend in popular music: the fusing of hardcore Hip-hop and heavy rock.

"I used to scratch the beginning of 'Walk This Way,' while Run would rap over it," said Jay.

When Jay started looping the record in the studio, intending to just use a piece, producer Rick Rubin had an idea: why not remake the whole record and bring Aerosmith in on the recording?

Rick Rubin later told author David Thigpen, "I remember how intrigued Steven [Tyler] and Joe [Perry] were watching Jay on the turntables manipulating their music. They were blown away."

The video is one of the most pivotal ever to grace MTV, and shows both groups practicing in adjacent rehearsal spaces, at first battling, then destroying the wall between them (symbolically tearing down the barrier between rock and rap), and then performing together on stage, ripping apart a racially mixed audience of Hip-hoppas and head-bangers.

In his excellent book *Jam Master Jay: The Heart of Hip-hop*, David Thigpen writes:

The sight of Jam Master Jay working the turntables while Run and DMC and Aerosmith's Steven Tyler danced made a whole new generation run to their turntables, and it opened up a door to a world of possibilities made real by blending rap with rock. Lots of cool kids still wanted to grow up to play guitar, but after Walk This Way first ran on MTV in 1986, many of them decided that becoming a DJ was cooler. And their role model was Jam Master Jay.

The album stayed on the charts for an incredible 73 weeks, and is credited with spawning nu-metal and rock–rap, and planting the idea that a DJ could either replace the band or play *in* the band.

As if this wasn't enough influence to inflict on pop culture, Jam Master Jay became co-founder, along with businessman Rob Principe and writer Reg E. Gaines, of Scratch Media Productions and the Scratch DJ Academy in New York City. Serving as dean of the Scratch DJ Academy, Jay was in charge of writing, developing, and teaching the class curriculum (Figure 4.5).

As an academic and a fellow teacher of DJing, I had the opportunity, in August 2002, to meet with Jay and discuss the possibility of future college-level accreditation of the Scratch DJ Academy's curriculum—one of the goals of the academy's founders. Jay also enthusiastically described his current tour with Run–DMC, Aerosmith, and Kid Rock, acting out their encore collaboration on *Walk This Way*, including a play-by-play account of his show-stopping turntable manipulations.

A few weeks later, on October 30, at 7:30 PM, Jay was shot dead by a masked gunman who barged into his recording studio in Queens. In the days that followed, the studio

Fig. 4.5. Jam Master jay demonstrates a turntable technique as dean of the Scratch DJ Academy.

became a makeshift shrine. Legions of admirers left candles, flowers, pictures, and even some turntables and a white pair of Adidas.

One homemade sign read, "Now God Has A DJ."

DJ Jazzy Jeff

Down the road 90 miles from New York, Hip-hop was catching fire in the city of brotherly love. Philadelphia DJs Spinbad, Cash Money, and Jeffrey Townes (AKA Jazzy Jeff) were perfecting the transformer scratch, adding rapid-fire 16th notes to the lexicon of turntable techniques.

When Jeff showed up at the New Music Seminar DJ Battle for World Supremacy in 1986, his reputation had preceded him. When he wiped out the competition, it was clear that Hip-hop was no longer just a New York thing.

Landing a record deal along with his MC, the Fresh Prince, the duo recorded *Rock the House* in 1987, which featured lots of Jeff on the decks, as well as the MTV hit "Girls Ain't Nothing But Trouble."

The follow-up record, *He's the DJ, I'm the Rapper* sold more than three million copies, won the first-ever Grammy award for rap music, and helped make Hip-hop accessible to suburban white kids. The record also contained some awesome transforming and chirping by Jeff, and while they don't always like to admit it, many cutting-edge DJs learned a lot from mainstream Jazzy Jeff and the Fresh Prince records.

The Fresh Prince turned out to be nice guy Will Smith, who had a TV and film career waiting for him, and gangsta rap's darker and tougher propensities gave white suburban teenagers a much more potent way to scare their parents than "Parents Just Don't Understand" from *He's the DJ, I'm the Rapper*.

Jazzy Jeff is still in Philly, where he runs A Touch of Jazz, the music production facility he founded in 1990.

Old-School Legacy

The Hip-hop DJ changed what it meant to be a DJ, and while some of the pioneers take exception to the term, they were the first turntablists. Just as the guitar developed by leaps and bounds in Spain in the late 1800s, the Bronx and other urban areas of the USA in the latter part of the 1900s provided the perfect setting for the rapid metamorphosis of the art of the DJ.

While the seven DJs profiled here are perhaps the most influential of Hip-hop's first decade, they represent the tip of an iceberg. The concepts and techniques that Herc, Flash, and Bambaataa put into motion have been imitated, mutilated, confiscated, and reconstituted by thousands of DJs. This very change is the true legacy of the Hip-hop DJ.

4.1

Rob Swift: Eloquent X-ecutioner

Rob Swift has been busy. He did the scratching honors on *Future 2 Future*, Herbie Hancock's long overdue follow-up to *Future Shock*, the album that spawned the single "Rockit." There's even a track on *Future 2 Future* called "Rob Swift," a turntable feature that Swift fills with musicianship, humor, and panache.

Rob's a former member of the now defunct DJ crew, the X-ecutioners, along with Roc Raida, Total Eclipse, and other influential DJs. Their album, *Built from Scratch*, sold 66,000 copies in its first week, debuting at number 15 on the *Billboard* album charts. Appearances on *David Letterman* and *Last Call with Carson Daly* helped make *Built from Scratch* the first gold record by a Hip-hop DJ crew.

Fig. 4.1.1. Rob Swift performing with fellow former X-ecutioners Roc Raida (closest) and Total Eclipse.

Their historic mix-CD, *Scratchology*, chronicles the evolution of the art of the scratch, including pioneering tracks like Grandwizard Theodore's "Military Cut," "The Adventures of Grandmaster Flash on the Wheels of Steel," and even Grandmixer DXT's scratching on "Rockit." These are rounded out by tracks by Davey DMX, Cash Money, QBert, D-Styles, and the X-ecutioners themselves.

Rob has been keeping a high profile in other ways, too.

It was Rob and Shortcut trading licks in the Gap's instant classic "boys who scratch" commercial.

Then there were his scratch lessons on MTV. And the gig with Quincy Jones, Herbie, and Branford Marsalis at the World Economic Forum. His scratching also appears on the new Blue Man Group record.

While the response to Rob's solo record, *The Ablist*, was overwhelmingly positive, his second solo record, *Sound Element*, has been hailed by some critics as the most musical record ever by a solo turntablist. We started our conversation talking about the music on *Sound Element*.

It's great to hear someone approach an album from the point of view of a musician who happens to be a turntablist. For instance, what can you tell me about the "Salsa Scratch" with Bob James?

I did "Salsa Scratch" to help people understand the depth of what you can do with a turntable. And at the same time, it was a chance for me to do a song that paid homage to where I'm from, Colombia and South America, and to my dad. His skills as a Spanish DJ, playing out at Spanish parties, playing salsa and merengue and stuff like that, is what inspired my brother and me to become DJs.

It's funny because out of all the songs on the album, that's the one that everyone comments on. The first words out of their mouths are "Salsa Scratch."

It's different kind of record for a turntablist to make. How did you come together with Bob James?

I grew up listening to Bob James. My brother John, who I credit for teaching me how to DJ, had a really extensive Bob James collection. Bob James is a really important figure in Hip-hop; a lot of his music has been sampled.

If you watch the movie *Wild Style*, you see Grandmaster Flash cutting up *Mardi Gras*, which is a classic Bob James record.

Peter Piper is a sample of that same exact song.

I grew up listening to him and scratching his records up, so the connection basically started from there. He's part of the reason I always considered myself a jazzy type of DJ—a DJ that would use a jazz influence in Hip-hop, along with Herbie Hancock.

When I was recording *The Ablist*, my first solo album, I recorded a song called "Fusion Beats," and in that song, a friend of mine is playing keyboards, and I'm on the turntables, acting like the drummer. That song caught the attention of Milan, a good friend of Bob James, and somehow he got my number and called me and told me how happy he was to see a DJ taking that route.

Milan said, "Rob man, I really like your work on *The Ablist*, and I actually know Bob James, and would like to, in the future, get you guys linked up."

At the time, I didn't believe him. I thought he was just reaching out to give me some props for how I went about recording *The Ablist* album, because there were a lot of live instruments on it. At the time, you weren't hearing that in a lot of DJ music.

A year later, Milan called me and said, "Look, I'm working on a project with Bob; I can really hear you on it. I really want you guys to meet up, and I have this session in Manhattan."

That was really cool on his part, man, because I was able to work on some projects for Bob on some albums of his.

I asked Bob to be a part of "Salsa Scratch" because one common instrument in most Spanish music is the piano.

Fig. 4.1.2. Rob Swift in a solo set.

And I didn't really know any piano players, so I asked him to be a part of the song, and he said yes, without flinching. So, that's how that whole thing came about.

How does Bob feel about Hip-hop and the whole turntablist movement? He seemed rather bemused by the whole thing a few years ago. Has his view on DJs evolved?

I think it's probably evolved because we've been working together. I think I can safely say that before we got together, he'd never really worked with a Hip-hop artist. He's only worked with other actual musicians—you know, "traditional" instruments.

When he hooked up with me and understood that there are artists like myself that do have an honest respect for jazz and all other types of music, and he saw how I really wanted to learn and gain more experience to help what I'm doing, I think it impressed him, and I think that naturally must've helped his outlook toward Hip-hop grow.

Although he wasn't familiar with what I did as a DJ, he jumped right into it, and it turned out to really benefit the both of us. Now, we're having fun trying to discover these new ways of working together.

I'm doing some gigs in a couple months with Bob James at a Blue Note [jazz club] in Japan: four nights, two shows a night. It'll be him and a sax player, and me on the turntables. It's going to be really interesting.

A jazz trio.

Exactly.

So, take me back a little bit. Tell me a little bit about your background.

I grew up here in Jackson Heights, Queens. My dad was a DJ. On the weekends, he would go through his records and try to figure out what he wanted to play at the next party he was doing. He'd take me with him to these parties, and I'd help him carry records, and sit there and watch him make people dance, you know?

My older brother, John, started DJing. Days off from school, my dad would go to work, all my brother's friends would come over and bring their parents' records, and they'd just jam, and make tapes, and have fun, and stuff like that. Naturally, I followed in my brother's and dad's footsteps. I wanted to DJ as well.

When I turned 12, I asked my brother to teach me how to DJ, because while my dad was my first example of what a DJ was, I wasn't interested in just playing Spanish music at a birthday party or a wedding. I wanted to learn how to scratch, which is what my brother was doing. So, he started giving me lessons on my dad's equipment.

I just fell in love with it, man, 'cause it was a way for me to be creative and a way for me to kind of salvage an identity for myself in the neighborhood.

Like, if you want a cool mix tape, or if you want a real cool DJ to DJ your party, you can go to Rob.

Between 12 and 18, I started learning about DJs from other parts of the nation, like DJ Jazzy Jeff, Cash Money, and DJ Aladdin, and all these amazing DJs that helped progress what Grandmaster Flash, Grandwizard Theodore, and Grandmaster DXT created. I saw that the art was really growing.

At 18, I started thinking, "Man, I really want to be more than a DJ who's just going to make a mark in the neighborhood. I want to make a mark in the whole DJ world."

I started to practice harder, and I really started to try and learn what it was that all these DJs were doing, how they were doing all the stuff they were doing, and why it sounded the way it did. I started studying the art form, as opposed to just doing parties and stuff.

How did you go about studying what other DJs were doing?

Audiotapes of DJ competitions. My brother's friends would have these audiotapes of, like, Cash Money battling at the World DMC finals in London. You couldn't really get a videotape of a DJ like a Cash Money or an Aladdin as easy as you can now.

So, I was forced to listen to what these guys were doing and try to figure out what they were doing with these turntables. A lot of times, I would get lucky and be able to figure out what they were doing. I started copying what these guys were doing at first, doing a carbon copy of a Cash Money routine, or a carbon copy of a Jazzy Jeff routine. I'd go get the records they were using, and figure out how to do exactly what they were doing on those records.

I'd invite friends over and do routines that they'd heard Cash do, right in front of them.

About a year later, I met up with a real pivotal person in my career named Dr. Butcher from Corona, Queens. He was light years ahead of me, but I didn't know about this guy. He ended up really changing my life as a DJ and my perception of what a DJ is.

Actually, a funny story, a friend of mine named Marcelo had a friend who knew Dr. Butcher, and my friend was bragging about me saying, "My friend Rob, he can beat any DJ in the neighborhood!"

His friend was like, "You're mistaken! This guy Dr. Butcher will eat your friend up!"

Marcelo got so mad he called me, "Yo, this guy says Dr. Butcher is better than you, man, why don't you get on the phone and tell him that he's joking!"

This guy got on the phone and I'm like, "Yo, bring your friend Dr. Butcher over to my house anytime, and I'll take him out, no problem."

That was that. I was like, "Dr. Butcher, who the hell is that?"

Fast-forward six months later, my friend Juju played a trick on me. He knew that I felt I was the best DJ in the neighborhood. Juju called me one day from Dr. Butcher's house, and he

Fig. 4.1.3. Backstage, Rob Swift warms up his fader fingers before a show.

was like "Yo, Rob! I just got these new turntables, and I started practicing! I got this new routine I been practicing for the past three days, I want you to hear it!"

So he had Dr. Butcher do this routine for me over the phone and my jaw just dropped. I couldn't believe that Juju learned how to do all this amazing stuff in three days.

The stuff he was doing sounded way better than the stuff I had been doing for the past six years! I hung up the phone, and I was like, "Wow, I better start practicing . . . ."

He called me back 10 minutes later and said, " I was just playing a joke on you man, that was Dr. Butcher!"

And I was like, "Word?! That was Dr. Butcher?"

What happened six months before, with that kid, totally jumped into my mind. I thought, "This guy's really good, he IS better than me."

That weekend, I met Dr. Butcher. Juju was nice enough to take me to his house. He turned out to be a really nice guy, and I think that when he met me, he saw something in me that made him comfortable enough to take me under his wing and show me the ropes, and teach me everything he knew about DJing.

So, he became your mentor, in a way?

Yeah, he's real spiritual, really into God and stuff like that, and the cool thing about him was, a lot of times, he would just sit me down and talk to me, and that in itself was like practice. He would tell me to believe in myself and give me books to read about self-motivation, and he just had me thinking on a whole 'nother level.

He told me, "Rob, you're good at doing someone else's routine, but you have to find yourself, and find how to give more of yourself to the turntables when you perform, 'cause you don't wanna just be someone else up there. You want to be you."

He showed me the ropes, and for a year straight, I would go to his house Monday through Saturday. Every day, 6 o'clock. He would get off of work at 5, and 6 o'clock, I would meet him just as he got home. I practiced with him till 8:30, 9 o'clock, every single day.

And the thing about it was he had a girlfriend and his girlfriend would come over and a lot of times she'd be like, "Why is this guy over here every single day at 6 o'clock?"

I didn't care. I just wanted to learn how to be the best DJ that I could be, and through his good-hearted self, I know a lot of those times he would rather have spent time with his girlfriend, but he never made me feel like I was interrupting or I was in the way. He would always sit with me and practice.

So, I feel he was the person that set everything in motion for me to really start believing that I could be better than a lot of the DJs that I looked up to—possibly even be better than him.

Another cool thing about him is that as much as he taught me, to this day, when you talk to him, he always figures out ways not to trip—not to take the credit for it. It's really cool. He'll say, "Nah, Rob, it was you. You really did it on your own. I probably helped you out here or there, or maybe said something, but you really did it on your own."

And to a degree, it was my aggression, you know. That I did make the effort to go out there and learn, and I was so intense, that to a degree, I understand what he's saying. But at the same time, if he wasn't open and didn't teach me a lot of things that he did, I can honestly say that I wouldn't be where I am now. But that's just the way he is—to not take

Fig. 4.1.4. Swift supports an MC by pitching a melody and beat scratching.

credit for it. And that's basically what propelled all the stuff that happened, as far as me entering battles and making a name for myself.

The cool thing about Drew, his real name is Drew, he was doing things that I had never seen or heard anyone do. It seemed like he could've been just as popular as Cash Money or Aladdin, but he never took it upon himself to go out there and make a name for himself. He always just kept it in his bedroom. That's something that always kind of confused me. I never understood why he didn't want to go out and battle other DJs and show people that he's just as dope. I guess it was just more a thing where it was a creative thing for him to just do.

What are some of the specific things he taught you?

The back scrape was a scratch that I had never really heard people do. He taught me how to do that. He taught me different techniques for beat juggling, like how to make beats sound like they're going in reverse—like they're going backwards. Back then, one of the popular beat jugglers was Steve D, who basically invented beat juggling. But Dr. Butcher was doing his own form of beat juggling. Not only would he make it sound like he was making his own beats with a record, but he would make it sound like he was making it go backwards, or in reverse.

How would he do that?

He'd take his finger and spin the record backwards in a juggle routine, but you had to do that in a certain rhythm. Stuff like that, I learned, and I applied to routines. Now, when people see me do them, they'll be like, "Aw yeah, all hail to Rob Swift!" But they don't know this is an influence from Dr. Butcher. So any chance I get to talk about him, I do, because a lot of what you see in me is stuff that I saw him do.

What was next?

From meeting Dr. Butcher, I got way better, and I started getting the confidence to enter competitions.

I know you were in the 1991 East Coast DMC competition . . .

Actually, I placed third. That was my first battle, and met Steve D there, and he's the one that basically inducted me into the X-men. And, a year later, I entered the 1992 East Coast DMC, and I won that. It was a good feeling, knowing that a year later, I won the battle I placed third in. That was the start of people knowing who I was, and people looking forward to seeing me in the next battle, like, "Whoa, what's Rob gonna do?" By then, I had already gained all the confidence I needed to get up there and do what I needed to do as a DJ.

Fig. 4.1.5. The X-ecutioners on the Scratch tour.

How much do you practice?

Nowadays, I probably practice an hour, maybe two hours a day, if I'm lucky. It's just so hard now to practice. Working on remixes for people, traveling, there's so many more responsibilities that I have now that I didn't when I was just coming up. I don't spend as much time practicing as I would like to, but even on a hectic day, if I didn't get any practicing done, I'll squeeze in a half an hour to stay loose.

Back in the days when you were coming up and then competing, what kind of practice time did you put in?

Six or seven hours in a day, and I'd go at it every day. No lie. And I miss those days, man.

What are your goals for the future?

My dream is to walk on stage at a Grammy awards ceremony and accept an award for a turntable album.

That's my ultimate dream. I figure, if that happens, it's going to basically reflect all the years of hard work that DJs like Grandmaster Flash and Grandwizard Theodore went through, and all the doors that they opened.

For me, no one can beat the feeling of being able to sit down with Herbie Hancock and work, or being able to sit down with someone like Bob James. Nothing beats the feeling of a fan coming up to you with an album and asking you to sign it. Nothing compares to that, no word, nothing.

But I feel like, symbolically, an artist like the X-ecutioners, an artist like QBert, Mixmaster Mike—with all of these incredible DJs, someone needs to win a Grammy. I think that is my ultimate goal, and that is what I'm working toward. Not because of it being an award, but I think it would symbolize that the movement has finally made it. You know what I mean?

I know exactly what you mean.

4.2

DJ Shadow: The Berklee Seminar

While growing up in the small college town of Davis, California, Josh Davis was about as far from the New York City Hip-hop scene as anyone could get. Somehow, Josh (AKA DJ Shadow) turned out to be a Hip-hop DJ.

Shadow is considered one of the pioneers of sample-based musical composition. James Lavelle was the first to give Davis a chance when he was starting up Mo'Wax Records in the UK. After many singles, Shadow dropped his first full-length album, *Entroducing*, which saw major popularity in the UK, Japan, and eventually the USA.

Shadow provided the music for the powerful documentary *Dark Days*, a film about homeless life in New York City subways. The music was largely made up of songs from *Entroducing* and *Psyence Fiction*, but also included the *Dark Days Theme*, which was released as a single in 2000.

He also appears prominently in the Sundance Award-winning Doug Pray film *Scratch*.

In *Scratch*, Shadow gives the audience a glimpse into his rigorous application of the Hip-hop tradition of *digging*, searching for records everywhere and anywhere.

Producing is another aspect of Shadow's career. He's played a major role in the production of records from Blackalicious, Latyrx, Lateef, and Lyrics Born on the SoleSides Crew label.

As a live DJ, Shadow was one of the first to embrace video DJing off of DVDs with the Pioneer DVJ. His current stage set-up takes video DJing further, with nine video screens presenting Shadow in silhouette.

Since *Entroducing*, Shadow's output has included *The Private Press* on MCA records, and *The Outsider* on Universal, which features collaborations with singers, rappers, and live instruments, in addition to Shadows virtuosity on the sampler and newly honed ProTools chops.

What follows are the comments that DJ Shadow conferred upon a packed recital hall at the Berklee College of Music before the release of *The Private Press* on MCA records, and represents an exciting moment in time for an artist who's artistic sensibilities and use of technology is constantly evolving. Full of his unique vision and philosophy of music and culture, this exchange of ideas shines new light onto one of Hip-hop's most notorious shadows.

The first music that I remember pursuing on my own was music that was very technologically advanced.

In 1979, Lips Incorporated had a record called *Funky Town*. I liked the robot voices and the laser sounds because it sounded like *Star Wars*.

My mom was always very suspicious of anything mainstream. She was always knocking anything that was too "show-bizzy." In 1980, the stuff I was subjected to was music like Eddie Money and Eddie Rabbit. So, that's what was going on, and I was like, "Man, I know there's gotta be something else!"

Fig 4.2.1 and Fig. 4.2.2. DJ Shadow tells about his journey from DJ to producer and recording artist at his seminar at Berklee College of Music.

Then I heard Devo. That was the first thing I ever spent my own money on. That and *Rapper's Delight* were the first rap records I ever heard. It still sounded kind of like disco to me, especially *Rapper's Delight*, which came out in 1979. That was considered to be the first rap record.

In 1982, my life changed when I heard *The Message* by Grandmaster Flash and the Furious Five. What blew me away initially wasn't the music; it was the raw delivery of the lyrics. It was more potent than anything I had ever heard musically.

It wasn't poetry; it was just reality. It was a very rugged song about a culture that I didn't know anything about, and about a city that I didn't know anything about: New York City.

Somewhere, I still have a tape. I used to lean over and tape the music that I liked off of a little AM clock radio. We're not talkin' about hi-fi here. I would just press record, and my parents would come in and be like, "What is this?"

I was goin', "Shhhhh, shhh, shh!"

A few weeks later, I heard *Planet Rock* by Afrika Bambaataa. These songs were "contemporary urban." Historically, a couple of years later, around 1985, you had the whole break dancing phenomena, and Hip-hop really blew up on a wide scale. It was a media fad, and a lot of hype. You had movies like *Breakin'* and *Beat Street*, and that took Hip-hop to a nationwide level.

The down side was that lots of people thought it had all been cooked up by the media. In 1985, everybody thought that Hip-hop, break dancing, graffiti, and everything associated with Hip-hop culture was just dead.

That was definitely a bad year for Hip-hop music because a lot of artists sort of internalized that attitude and thought, "Well, I guess we're just supposed to get paid real quick and get out."

Then 1986 came around. People used to come up to me and be like, "Oh yeah, you're the break dance kid. You're the rap kid. Why are you listening to that?" It was music that was absolutely despised by every mainstream media outlet from *Rolling Stone* to MTV. They didn't realize that there was a much broader cultural context.

Now, I'm living in the major-label reality of the record world, and it's ill. Some of you might want to know how I ended up here.

I bought my first turntable in 1984, and I was imitating all my heroes.

I was imitating scratches that I heard on records like *One for the Treble* by Davie D or any of the early Run–DMC stuff with Jam Master Jay.

Any record that had a lot of scratching in it, I was trying to do.

In 1987, I started playing mixes on a college radio station in town [Davis, CA]. I was a sophomore in high school and my first mentor was a guy named Oris. His handle was "Big O, the Ultimate Gigolo." He had the first and longest-running rap show on KDDS. He was the first person to say, "Man, how are you hearin' about all these rap records? I'm doin' a rap show, and I've never even heard of this stuff."

So, I started playing these mixes and bumped into a rapper named Paris. He was from San Francisco and was going to school there. When he got signed, he blew up. A couple of years later, after he graduated, we hooked up, and I started doing some production work for him.

Eventually, I just started sending tapes out. There's this station in San Francisco called KMEL. It was a very influential urban station. This station, at the time, was breaking records like Naughty By Nature's *OPP*, Chub Rock, and Main Source's *Lookin' at the Front Door*.

I called up the program director there, and said, "You need to hire me. Nobody else is doing all Hip-hop mixes."

He said, "Well, why don't you come up and tell my why?" I did, and he said, "Well, you're pretty funny. I'm gonna give you a slot."

By now, I had started college. I still didn't really know what I wanted to do. So, I sent some tapes around. My confidence was up. At that time, if you were doing mix shows on KMEL, you were getting jocked by every record label person who wanted their product played on the air. I just used to get bucket loads of promos from all the labels, and got to know all the A&R people. I said, "Well, check this out. I've been sending you my mixes so you know that your stuff is being played on the air, and now I've got some beats I want you to check out."

We're talking about labels like Profile, Wild Pitch, Big Beat—all pretty reputable labels. All of them wanted to give me work, but would say things like, "Why do you have to sample this weird stuff? Why can't you take familiar samples and loop them around?"

That's what everybody wanted me to do, and I was really depressed about it. But, one man gave me a shot. He worked for a Hip-hop magazine called *The Source*. I would send him a tape and say, "I think you should do this. This would be a good thing to put out."

He would kind of laugh, and say, "Okay, I think I can get you three grand for this mix." I would hang up the phone and be like, "Very good!" That's when I thought, "This might work out!"

Persistence is just the ultimate key, and also having confidence in the fact that you are presenting something different. If you're just trying to be yourself and offer music you really love, then somebody will eventually pick up on your passion and reciprocate that. That's what I think happened to me.

We had a show, one time in 1993, and we were sitting there putting the labels on the records in the parking lot before the gig. We had 500 copies just to give out. And like I say: persistence. Record after record after record, never expecting anything. You never wanna put all your eggs in one basket.

You just have to put out everything you do with the thought that, "I'll be lucky if this gets any attention whatsoever."

I think, if I had my way, I'd be at home all the time, just working out of my house. But, people like to see that you can come out there and project yourself and not be afraid to play in front of people.

My next big break was when I got a call from James Lavelle at a label called Mo' Wax. He said, "Hey, would you like to do something really different?" And I said, "Yes, please!" All these other rap labels just wanted me to recycle. I did that, and it caught on in England. I worked that angle and stayed out there for a long time.

Next thing I knew, I was putting out *Endtroducing*. I think if you just do things for the money, you're never going to make it anyway, so I like to do things that are going to be fun for me.

Where did you get your name?

I chose the name Shadow because, at the time in late 1989 and early 1990, a lot of established Hip-hop producers were starting to step out from behind the boards and try to make their own name. I just thought that a producer's role should be in the back. I identify more with directors than actors. Directors can call all the shots, and they can still walk down the street and nobody knows who they are. I just liked that.

Did you ever consider putting out a mix CD like Dan the Automator?

Yeah, actually, I have done something before. I did something called *Brain Freeze*. My man Cut Chemist from Jurassic Five and I wanted to do something unique because we had never DJed together before.

We decided to do a mix routine using all 45s. We performed it in San Francisco, and luckily, I taped our last rehearsal before we went out to the club, and that's what ended up being *Brain Freeze*.

People say, "Well, what's the big deal with mixing 45s?" But if you've ever tried, then you know what I mean. They're just a lot more temperamental, and they jump a lot easier.

How did you make the selections for your *Brain Freeze* record? I think that was just some of the hippest, most different-sounding stuff, and I was just wondering if you had that in mind before you started on the record.

One of the reasons Cut Chemist wanted to do this with me was because—and I'm not tooting my own horn because he's got strengths that I couldn't match—but he knew I was winning on the 45 side. And he just wanted to get inside the boxes and see who's in there. We just sort of dumped everything out on the table, and went, "Okay, we know we gotta play a certain amount of samples that people maybe don't know what the original source is. We gotta play a few things that are gonna get the floor going. We gotta play a few things that are gonna make people think." I think *Brain Freeze* was a good exercise for us to do, but I think *Product Placement* was a little craftier.

On *Endtroducing*, you had a track called *Why Hip-hop Sucks in '96* with the punch line being "It's the money." What are your current feelings about Hip-hop?

It's a complicated thing. In the mid-1990s, if you were making Hip-hop, you either belonged in the underground category or the commercial category. If you were making underground Hip-hop, you had to go to great lengths to articulate what time it was, and that you were down and weren't going to try to exploit the cultural aspects of Hip-hop. I think it's real cool right now.

At the time, I was frustrated—not only with the commercialism of the mainstream stuff, but also this real incestuous and closed-minded attitude of a lot of people in underground Hip-hop.

People were trying to keep a lot of aspects of the culture alive. I feel that Hip-hop music—rap music—is real strong right now.

Could you talk about *Dark Days* and the film industry versus the music industry?

Dark Days was a documentary that I wrote the music for. It's the only film work I've ever really enjoyed. All I can tell you, from my experience in making film, is that the film industry can be even more obnoxious than the music industry. There's more money at stake, and people are more neurotic, and oftentimes, music is the very last thing the director thinks of. They think visually.

A lot of them do not think musically, especially if they're trying to make a Hollywood blockbuster. Music is just like, open up the latest copy of *Rolling Stone* and pick out four hot faces and ask them to do something for a soundtrack.

If my name is attached to something I need be able to put my weight behind it. *Dark Days* was the first time anybody came along and said, "This movie is at least a year

Fig. 4.2.3. DJ Shadow addresses a full house in the David Friend Recital Hall at Berklee College of Music.

away from being finished. I want you to really check it out." And when I saw it, I was blown away.

It's about homeless people living in the subways in New York, and the movie was also made by the subjects.

It's very fascinating and life affirming, and all the things that you hope a good movie or a record would be.

Is it a pain clearing all of the samples you use? Is there a lot of paperwork involved? How do you go about it?

Sample clearance is difficult, in certain cases.

My personal philosophy is that my music is a collage medium.

I look at it and go, "Yeah, that's a new piece of work." You didn't just take one giant image and put it in a new frame, like a lot of people who've gotten in trouble with samples were trying to do. I feel like what I do is a lot different.

Having said that, if I do sample something substantial, I don't have a problem trying to clear it, but I try and use things that are really off the beaten path. If I was just sampling James Brown and Sly Stone, it's easy. They have a whole business in place that takes care of sample clearances.

One sample on my new record is a whole vocal track, which of course I'm going to clear, but the record I took it from says "Number 321 out of 500." There's no information on it whatsoever, so we're just stuck. All I can do at that point is put it out, and hope that everything works out.

Was *Private Press* more software based, or was it all done with your Akai MPC sampler?

Personally, I've always been afraid of embracing too much technology too quickly, because I don't like to spend my work time reading manuals. I'd rather just plug in and go.

So, in that respect, I do use ProTools, but I still program in the MPC.

I have two MIDI-ed together now, which is actually something that the Bomb Squad, Public Enemy's producer, used to do in the late 1980s. They had two or three SP1200 drum machines MIDI-ed together. That is how they were able to create, what is, to me, some of the most amazing sample-based work ever.

What kind of Akai MPC are you using these days?

In 1996, I bought the 3000 and shortly thereafter, the 2000 came out, but I didn't like working with the waveform. I liked working with numbers instead, so I stuck with my 3000. Eventually, I bought another 3000. Now I just use both MIDI-ed together, and I do all my sequencing in the MPC.

Do you find the more that technology becomes available and the easier it gets to use, that you use it more?

Always, in the past, I've tried to be very cognizant of not focusing too much on technology.

I like the music to be from the heart, not from the brain.

When I was working on this record, I decided to switch it up and do a song that's based entirely on a technical concept.

One song I wrote, called "Monosyllabic," was made entirely from one sample. Every single sound had to be extracted from the same two-bar loop. It was definitely the most labor-intensive thing I have ever done. And then to make it a little bit harder, I prohibited myself from using software programs to make crazy sounds. For example, reversing a small second of the sample, gating it, putting it through a Leslie cabinet, having a speaker hung in front of it, and then re-sampling that new sound. It's very unlike anything I've ever done.

How do you decide what direction you are going to take a certain track or sample? Is it an experimental thing, trying different things out?

Yeah, plus lots of times, the nature of the sample will dictate what it should be. Some things just sound like they need to be treated in a certain way. I try to let the song and the sample dictate that, rather than saying, "I'm gonna run everything through my favorite piece of gear."

But, like I say, unlike a lot of producers—and I sometimes consider this a fault of mine—I'm just not up on all the latest gear. I spend more time taking in new music for inspiration, rather than scouring the music stores for the latest, greatest program.

Do you ever do live overdubs? Say, a live bass line?

I like when producers mix live instrumentation with samples. Organized Noise and Outkast, for example, do it.

For me personally, though, the sampler is my instrument of choice.

It's the instrument I most identify myself with. I consider it a challenge to just constantly try and motivate my ears to do new things with samples. For that reason, I resist, whenever possible.

For this album, it's 100 percent samples again, and I didn't actually set out to do that. A lot of times, you'd have a record that people would find amazing that was just a four-bar funk loop with a sax playing over it for four minutes. I just thought, "Man, this is lazy." That's why, when I first started doing press, I was very vocal about no live stuff.

But I listen to music that is made with live instruments every single day. I have nothing but respect for anybody that plays a live instrument. I took a couple of years of piano, and I just couldn't hack it.

Do you have any theory or harmony training, and do you use that in your music or is it more ear-based?

It's all ear-based. I've only now started to use samples harmonizing with each other—re-sampling in different intervals. I'm just kind of getting into stuff like that. Vocal soul or northern soul from like '65 to '68 and also some psychedelic stuff from like '67 or '68 is really good for understanding vocal harmonies.

When you look for samples on records, do you separate them by beats and bass lines or textures?

Yeah, on this album, I tried to get a lot more disciplined. I had a Post-it note system. I would just play through records and write, "Good break for something," and I'd stick it on there and put it on the shelf. But on only one out of ten records do I actually find something.

I look for unusual sounds, like strange studio accidents or the way something was mixed or effected. And then I separate the records into shelves; this will be this song's shelf, and this will be that song's shelf.

When you're working on a new track, what do you look for in your samples?

My favorite records to sample are those that are not great records but have moments of greatness.

I like to extract the sole moment of genius on an otherwise awful project.

And I don't want to sound real judgmental.

As far as I'm concerned, anybody who ever puts out a record deserves respect on any level. I honestly feel that, because it's just not easy. If it was easy, everybody would do it.

5

The Scratch DJ Revolution

"My first love? Boys who scratch," cooed the seductive Shannyn Sossamon in the classic Gap ad featuring Rob Swift and Melo-D trading licks over a slow groove. A lot of guys took notice.

Although many old-fogy DJs dismiss the very notion that scratching should be the focus of a DJ's career, no one can deny that there is now a generation of DJs who were drawn to the art form by a desire to scratch.

However, the term scratch DJ, while in wide colloquial use, is not even universally accepted.

"It was called cutting," Grandmaster Flash told Terry Gross on Fresh Air. "It's now called scratching, which is just like one part. It's almost like saying to a boxer, he's boxing, but now we're gonna call it 'right hook'. The right hook is only one area of a boxer's skill. Scratching is just one area of what this thing entails."

That hasn't kept the term from blowing up. Doug Prey's film, released in 2000, about the scene is called "Scratch." The school co-founded by Jam Master Jay in New York (which has now been franchised to Los Angeles) is called the "Scratch Academy." The two most established computer interfaces for DJs who want to use their turntable techniques to control audio files are called "Final Scratch" and "Serato Scratch."

If one DJ described another DJ with the words, "She's a great scratch DJ," chances are, everyone will know what's being said. But what exactly does this term imply?

"Before you learn how to scratch, you must first learn how to mix," says DJ Babu, the DJ credited with first coining the terms "turntablism" and "turntablist."

While many scratch DJs are excellent mixers, some confess to not having used more than one turntable for a long time. DJ QBert, who has taken scratching and especially "beat scratching" to entirely new levels, even developed a single turntable/mixer combo called the QFO, since most of his work is done on a single turntable (Figure 5.1).

The terms "turntablist" and "battle DJ" are also in wide use to describe DJs who are principally involved in manipulating vinyl, faders, and switches rather than mixing records together. A "turntablist" is someone who approaches the turntable as a musical instrument, as a guitarist or a pianist would approach his or her instrument. We'll delve more into this topic in the next chapter. The "battle DJ," on the other hand, is someone active on the battle scene, which has developed into a worldwide culture of its own.

Fig. 5.1. QBert playing the QFO, a turntable/mixer combo designed especially for scratch DJs.

The Battle Scene

The Battle DJ must be able to showcase their skills in short (often six minute) routines, which usually involve mixing, scratching, and beat juggling. Sometimes battle DJs will also go head to head in single elimination scratching rounds, reminiscent of the MC battles depicted in the Hollywood feature film "8 Mile." These battles are not all that dissimilar from "cutting sessions" engaged in by jazz musicians in Harlem in the first half of the 20th century; the pressure is high, and one mistake—a skipped needle, a sloppily executed technique, or a poorly conceived idea—can spell failure.

While there are scratch DJs who never battled, or barely ever did (Grandmixer DXT and Kid Koala for example), many, if not most, scratch DJs were inspired by seeing a video of a DJ competition along the way.

While the Zulu nation began sponsoring DJ competitions in the early days of Hip-hop, it was an English organization known as the DMC that first started widely distributing video tapes of its final rounds. This gave bedroom DJs from across the globe something to aspire to, along with a chance to see world's best DJs up close and steal their licks.

Originally conceived as a mixing contest, the DMC competition was turned upside down in 1986 when DJ Cheese introduced scratching in his routine, changing the course of the DMC battles forever. Since then, the DMC World DJ Championships have been won by likes of Cash Money, Cutmaster Swift, QBert and Mixmaster Mike, Roc Raider, A-Trak, DJ Craze, and I-Emerge.

The ITF, or International Turntablist Federation, was founded in 1996 by B-boy and San Francisco bay area promoter Alex Aquino. From the start, The ITF's competitions have included separate categories, including Scratching, Beat Juggling, Advancement Class, and a separate category for Turntablist Team/Bands. In recent years, an "X-perimental" category has been added, allowing DJs to use external effects, CD turntables, and laptop interfaces, instead of just the traditional bare-bones battle set-up of two turntables and a two-channel mixer which has been the battle standard for years.

Fig. 5.2. DJ Z-Trip performing on the Scratch tour.

The ITF competitions also boast an illustrious list of winners, including the Beat Junkies (J-Rocc, Babu, Rhettmatic, Melo-D), the Allies (Develop, Spiktacular, J-Smoke, Infamous), Total Eclipse, A-trak, Prime Cuts, Craze, I-Emerge, and Lil Jazz.

There are hundreds of other DJ battles taking place around the world, with sponsors as large as the retail chain Guitar Center, to neighborhood kids getting together to do battle in someone's basement.

The audience for most DJ battles is made up primarily of aspiring battle DJs and their friends. Attracting a mainstream audience to a DJ battle is a nut that DJs and promoters have to work hard to crack. Having judged quite a few battles, I must admit that, other than the finals where the best are going head to head, battles can sometimes become rather tedious.

This is not surprising—some scratch DJs are simply more musical and entertaining than others, and therein lies the keys to success.

Scratching Out a Career

Many famous scratch DJs, like QBert, Craze, and A-trak, earned their stripes in battle, and have parlayed their chops and the publicity surrounding their winnings into successful careers.

So, what are you likely to see if you go to check out a scratch DJ on tour? Well, that varies widely. DJ Craze spins drum 'n' bass for most of the night, and gets into scratching for only the last part of his set. QBert tends to play pretty short sets that include his virtuoso scratching over beats that he plays through the QFO's aux input off a mini-disc, as well as his incredible solo drum scratching. A-trak tours with Kanye West. Rob Swift likes to keep it fresh with guest MCs, DJs, and musicians sitting in. Kid Koala cuts up and mixes the most unlikely of records into a live collage that might please your mother, if she was open-minded.

DJ crews like the former Invisible Skratch Piklz and the X-ecutioners would mix things up in a variety of ways, from organized trades and body tricks, to forming a band where one DJ lays down a beat, another creates a bass line, another provides a contrapuntal rhythmic or melodic figure, and another solos over the top.

The inventive group Gunk Hole eschews the term crew in favor calling themselves a DJ band. This brings up the fine line between DJ and musician, a fine line which is examined in the next chapter, and is crossed so often that even having separate chapters on scratch DJs and DJs morphing into musicians has proved to be a precarious if not specious undertaking. Who to put in which chapter?

While all scratch DJs are not musicians, the truth is that all really *good* scratch DJs probably are. It is their very musicality that makes them interesting in the first place. And while you can say that a professional dancer is very musical without having to contend that they are indeed musicians in their own right, it is hard to extend the same logic to scratch DJs. We consider conductors to be musicians, and they, like DJs, are not playing an instrument, but organizing and expressing and reinterpreting the music of others by controlling the performances of musicians.

Scratch DJs take things one step further, by actually manipulating sound in a musical way; bridging the gap between musician and DJ with sonic collages that, at the very best, stand up as art.

5.1

DJ QBert's Perpetual Revolution

QBert's boundless imagination and insatiable curiosity have driven the art form of scratch DJing for over a decade.

Born Richard Quitevis, QBert was mentored by Mix Master Mike in the filipino Hip-hop DJ capital of Dailey City, California. While his DMC wins in the early 1990s (both on his own and with Mix Master Mike and Apollo) helped put QBert on the map, the folklore surrounding QBert's battle career since then has probably done just as much for the DMC.

QBert doesn't follow the rules of scratch DJing; he creates them.

Innovation is also the staple of his career, and for this, he has a secret weapon: his business partner, Yoga Frog. Playing Katzenberg to QBert's Spielberg, Yoga Frog's business acumen helped Q's transition from champion to phenomenon resemble similar transformations by the likes of Tony Hawk or even Arnold Schwarzenegger.

Fig. 5.1.1. QBert playing the QFO, the instrument he designed, which combines a turntable and mixer into a self-contained musical instrument.

The two have partnered to create Thud Rumble, an innovative company involved in producing everything from scratch records and hybrid slip mats to instructional DVDs. QBert and Yoga Frog seem to have an uncanny ability to sense where the culture is headed before anyone else gets a clue.

Turntable TV was reality television before such a thing even existed in the USA. Basically home movies by QBert and the Skratch Piklz sold on VHS, *Turntable TV* gave fans a way of connecting to the heroes of the scratch world. QBert, Flare, Yoga Frog, and company (as well as their alter-ego puppets and characters Lamb Chop, DJ Bushwhacks, and Scratchy Seal) became as accessible though *Turntable TV* as the Osbornes became through MTV several years later.

Every scratch DJ on the planet probably has at least one Dirt Style record, the label associated with QBert and the Piklz.

Thud Rumble's Skratchcon 2000 brought together a who's who of scratch DJs for a weekend of educational seminars and performances in San Francisco.

QBert's solo album, *Wave Twisters*, was adapted into an animated movie. He's essentially the star of the movie *Scratch* and headlined the movie's tour throughout the world.

When QBert is not touring, he is teaching anyone who wants to learn how to scratch. When secrecy about technique and records was common in Hip-hop and the battle DJ world, QBert and the Skratch Piklz were among the first to actively and willingly show their techniques to anyone who was interested.

QBert and Thud Rumble have created a DVD series titled *Do-It-Yourself*, in which Q gives lessons on scratching techniques, shows off his comic acting chops, and sets up battles between you, the viewer, and some of the world's most bizarre DJs.

Not long ago, QBert and I talked all night about music, scratching, how to practice, and what his perspective is on everything from his career to other DJs. Here are some highlights.

What are the pros and cons of using the "Ahhh" and "Fresh" samples to the extent that the whole turntable community uses them?

I think it's a standard that you can base the skill of a DJ on, by how he messes with those sounds. For example, if everyone used a certain type of clay for sculptures as a standard, you could see the skill level of each sculptor, based on what they could do with that clay. With DJing, when you hear someone use that sound, because it's a standard, you get to see the personality of each DJ.

Do you still go to battles?

I do go to them. I get inspiration from battles. Everyone has their own interpretation of battling. For me, it's just continuing the art of it.

There are guys that do a lot of the stuff that older DJs did before, but then, there are a few—like DJ Craze or A-trak—that do something more advanced than what was done previously (Figure 5.1.2).

Fig. 5.1.2. A faithful legion of fans show up to hang on QBert's every scratch.

Some DJs develop amazing skills early on in their career, but then they just keep doing the same stuff. On the other hand, you seem to keep evolving and growing. Where does that come from?

When I was a kid, I used to see these break-dancers, and I'd always say, "God, he got so much doper!" I'd come back the next day, "Man, he's so fresh." Come back after a couple of weeks, "Man, look at the new moves he got! That's how to do it." Then all of the sudden, he stopped. Why did he stop? He could've been so much better. Then I thought, what if I just never stopped? Let's see what happens.

What's the most unusual thing you do when you practice?

I like to have a trainer around me when I practice. I like to have someone to coach me. 'Cause I can get into my own world, and it sounds good to me, but to the regular guy, they're like, "Man, you're doing too much."

That helps me out; he'll tell me, "You're scratching too much … pause more … lay off here … do some more phrasing … mix it up!"

You do that a lot?

Yeah, Yoga Frog is just right next to me all the time. "Hey, do this, do that, do that."

Makes me realize what it sounds like to someone else. It's a kind of like a guy painting something and the average looker would say, "Hey, what the hell are you doing?" I like the outside opinion. Helps me a lot.

You mentioned phrasing and playing less. Seems like a lot of scratch DJs are just non-stop.

Yeah, sure, sometimes that's my problem. I do that a lot. I'm learning how to phrase.

It's like being lyrical. You put your rhymes in weird spots, and give it space for the listener to think, "Ah, that was a nice lick."

I'm learning that, if you watch a lot of my stuff, I'm always going for it, which is bad. Once in a while, in a mode or a phrase, yeah. But I need to learn to do that well.

Are there things that you do to warm up every day?

I love to stretch. I count to 60 and stretch every part of myself, my arms, my body, my back, my shoulders, and my chest; just to loosen up. I also drink a lot of water; it loosens you up.

Explain your practicing routine.

I always practice at least two hours a day.

I always start off slow, so that I can get the intricacies correct.

Then I move up to faster speeds as the time goes on. I start off with a slow beat, and then a medium beat, and then, of course, go to some electro stuff, some jungle kind of stuff, drum 'n' bass, get faster.

Then I'll start scratching some drums and then go to experimenting. I'll look at all my lists of scratches, and I'll just go, "Number 151 plus number 382." I'll put those scratches together, and then that just develops into some other kind of stuff.

Tell me about this list you grab from.

Oh, it's just a whole bunch of scratches and stuff. It just keeps going on forever. You can mix and match anything. Combine two things and the combination becomes a rudiment in itself.

What drives you to organize and write down your scratches?

I don't know, I don't think I'm as organized as I should be.

You've been responsible for giving names to a lot of techniques. What's the advantage to naming a scratch?

Having a name for a scratch gives it a picture, rather than just the "one-click combo to a three-click whatever." I'm sure that's a great way to think of it, but when you think of a name, like the "rabbit toucher," a picture gives it easier organization in the dream world, which is your mind, rather than the logical world.

How do you practice coordination?

That's a really big challenge for me. Coordination between your mind and your hands is really important. In my mind, I have all of these patterns, but my hands are doing other things.

Do you always have turntables set up in your hotel room when you're on the road?

Always. I did a DJ competition a long time ago in 1991, the World Competition, and I got beat by this guy DJ Dave. I don't know if you've seen that competition, in 1991? I didn't practice for three days, so I went on stage, cold. My practice was the battle; I was like, "Oh damn."

So, that taught me a big lesson: always practice before the show. Bring your turntables before the show. So ever since 1991, I have always brought my turntables and set up in the hotel room before the competition. Just so I can get warmed up.

Makes sense.

Normal routine.

You sound like you've spent some time transferring drum rudiments onto the turntable, have you?

I love rudiments. I had a drum rudiment book and the right hand would translate to a forward and the left would translate to a backward. I still do a lot of rudiments today. The drum rudiment book is mad.

Are you getting any more interested in melody playing now that the turntables slide +/− 50 percent?

Definitely. I'm learning to play bass guitar, so that's helping me too. I'll try something on that bass and say, "Ooo, that's cool!" I'll mess with the turntable, and try to get it on there.

I'll name some musicians; tell me what comes to mind.

Okay.

Jimi Hendrix

Freedom!

Miles Davis

Poetry.

Fig. 5.1.3. DJ QBert is constantly developing new right-hand techniques, and spends time in most shows scratching without the use of the fader.

Britney Spears

Popcorn music? [Laughs] Lollipop music?

DXT

Innovative.

Herbie Hancock

Jazz mentor. My friend DJ Disk and I were talking yesterday about hanging out with Herbie Hancock. Disk said that he has learned so much from him and has so much respect for Herbie Hancock.

DJ Swamp

That guy's a genius! He actually has this new video, *Worship the Robots*. We're making a turntable DVD and that will be on there. He's really getting into some visuals and other crazy stuff. That guy is very much in his own realm with being a DJ, with burning himself and all. That's some crazy stuff.

DJ Radar

Radar! Man, I learn so many drum patterns from that guy, he's a genius! He's doing that orchestra thing. He can read notes and stuff. Man, I wish I could do that. All these geniuses everywhere!

DJ Shadow?

>DJ Shadow is more of a producer kind of guy. I saw him at the museum the other day, and he was like, "Hey Q, we gotta hook up so you can show me some new scratches." He can scratch pretty well, but his main thing is his production. His beats are so beautiful.

Paul Oakenfold

>That is the, uh, house DJ? The number 1 house DJ?

What's your relationship to the dance music world?

>I just know that they have a much larger following than the scratch world. It's great that they're doing their thing, seeing a lot of monetary reward out of it. I give them a lot of props for that. They really deserve it.

The business side of what you and Ritchie [Yoga Frog] have been doing is certainly noteworthy.

>That's all Ritchie. That's all Yoga Frog. The mastermind of it all. There was a period when all he studied was what Bill Gates does and what the government does. What any top person does in their field, and he'll just incorporate that into the scratch business.

How did the Apple Macintosh commercial come about?

>Oh, that's the strangest thing. [Laughs] I hate myself on there. I was so nervous.

Tell me about your little toy turntable.

>Oh, they wanted me to bring something turntabling, scratching. I said, "Hell no, that's heavy." [Laughs] So, I just brought the little turntable, I thought that would be a little more interesting. Turns out that I really should've brought some real turntables, did something better on the commercial.

So, Tony Hawk was at this shoot?

>Yeah, he was right there! Tony Hawk and Kelly Slater. We were kicking it with Kelly Slater. He was showing us his home movies on his Mac, and we were outside in the parking lot having sandwiches and stuff. It was cool. Nice people.

You work both in regular mode and hamster mode. Could you talk about which techniques you do in both and why?

>I only juggle in regular mode, and I scratch in hamster mode.

>I scratch in reverse because when I got my first Radio Shack mixer, it didn't have a crossfader, so I would scratch up and down.

>So, when I would put my fader in reverse, it's more like an up and down fader, that's all.

Do you have any advice for people in terms of practicing crabs?

The concept is simple: you use your thumb to spring. And you rub your fingers across, kind of like you're snapping all your fingers.

Could you elaborate on the whole concept of "the Flare?"

For my generation, we were always accustomed to starting the scratch with the fader off. Then Flare showed us the Flare scratch around 1990. It was starting with the fader in reverse, starting it "on." It was just turning the whole world upside down.

So, I hear that you're moving to Hawaii?

Yep! I just bought a place in Hawaii. I still have my studio, the Octagon, in Daily City. I'm building a new studio in Hawaii. Ritchie and I are building this place in Hawaii called the Temple War-Plex, which is a place where artists can come and relax—kind of like a lounge type of club environment where DJs can cut, and then we can go upstairs and dance or whatever. B-boys can break, graph artists can commune, I don't know, it's just a place made for artists and stuff.

What's the Hip-hop scene like in Hawaii?

It's a really untapped world. A lot of great MCs and a lot of *great* B-boys. A lot of great DJs. I'd say that some champions are from Hawaii. I think it's just the whole environment, it's such a beautiful kind of place, and the weather is perfect every single day!

It's such a great place to practice. It's very inspiring to see the nature and the blue ocean, just scratch all day and look out the window just see God's beauty, how he created the world, kind of thing. Totally makes you feel one with the earth's energy!

What do you want to accomplish that you haven't yet?

There're plenty of things! I see the way these jazz musicians play, and I think, "Oh man! How do they do that?" My thing is just to be able to do stuff in one take and have it be clean and beautiful and interesting. It becomes their language.

It's so easy to talk with our mouth, and everything comes out perfect. To be able to do that with my hands, that's what I want to accomplish.

When was the last time you were totally blown away by a turntablist?

We have these DJ sessions at my house every few weeks, and everyone that comes in there brings new things. I get blown away by all the new things they come up with. All the new young cats have new visions of scratching. It's always such a treat to have them at the jam session.

What's your favorite new record?

Louie Armstrong's Hot Fives and Sevens recordings. That's my favorite stuff, right now. I guess I'm going back to that era.

Do you enjoy collaborating with other musicians?

> I do enjoy working with other musicians. What I really like about it is what they teach me. I will ask some musicians about melodies and what kind of emotion certain notes might bring out.

How do you feel about distributing music on the Internet?

> I'm doing that now. A lot of people can't afford music, and I feel that music should be free.
>
> I think that rewards come to the artist in different ways, not just money.
>
> I'm very spiritual, so if I give something out, whatever comes back to me is good. I don't care if it's money or if it's happiness in life.

When Hip-hop was first coming around, a lot of the DJs were very secretive about what their records and techniques. But you're totally the opposite.

> Exactly, because everyone has their own personality with it, and I just want to see what their personalities are like. Knowledge of what they're doing. I wish someone told me a lot of these things. I had to learn on my own.

Have you ever had to have a regular job?

> When I was very young, I did a lot of construction with my father. I was also a telemarketer.

What was your first memory of music when you were young?

> I was always waking up to classical music in the morning, and I liked it. My mom said I would be in her stomach and whenever she would play music, I would kick. Also, when I was a kid, I had a tape recorder. I would always listen for weird songs on the radio, and record those weird songs.

What about your mom, what does she do?

> She is a businesswoman. She was actually the Mary Kay cosmetics recruiting queen for the United States for four years in a row. She would go fly to Dallas, and they would crown her the queen of recruiting.

Pink Cadillac?

> Pink Cadillac, yeah. She had, like, three Pink Cadillacs.

Tell me a little about your record *Wave Twisters*. Did you put it together in ProTools?

> Yes, ProTools. I just did all the music on there and recorded a lot of it on ADAT at the time, too.
>
> Lately, I've really been thinking what it would be like if that stuff was never around? Look at jazz musicians; they never had any of this technology. Thelonious Monk said that he always used the first or second take and never the third, because that's when

Fig. 5.1.4. QBert and the author in Boston.

you lose the energy from the song. I want to be able to do that—just do everything in one take.

I love constructive criticism. From what you've seen, what should I be doing? How can I better myself, what do you think? I love honest opinions. It helps me so much.

Your whole spirit of improvization is jazz-infused. What I'd love to see you do is collaborate with some world-class jazz musicians to make a record that blows the roof off of people's preconceptions about the limitations of the turntable as an instrument. It would be a risky thing to do, but I think it would be amazing.

Wow, really? I'm willing to look into that.

I think it could be groundbreaking. Do you consider yourself a turntablist first, or a musician first?

I think that both are the same thing. I've always just called it "scratching." It's all art. It's just soul coming out.

5.2

Kid Koala: DJ as Storyteller

Kid Koala is a bit difficult to pin down. His book, *Nufonia Must Fall*, is a love story between an office girl and an out-of-work robot who carries around a record player. The book weighs in at 338 pages, and comes with a CD soundtrack consisting of Koala playing Wurlitzer electric piano and turntables. There is no dialog in the book and no lyrics on the CD.

Fig. 5.2.1. Kid Koala creating a sonic collage.

At a recent concert, the audience held its breath as he breathed new life into the vintage Henry Mancini track "Moon River," from the movie *Breakfast at Tiffany's*, by totally re-inventing the arrangement, using two copies of the vinyl record, with beat juggling chops worthy of a DMC champ, and the creativity of a world-class jazz soloist. During intermission, Koala handed out little pencils and bingo cards, then flashed hand-drawn *Nufonia Must Fall* images from his vintage slide projector until an audience member called out "Bingo!"

While most famous as a scratch DJ, Kid Koala (AKA Eric San) has a background in classical piano and early childhood education.

He has been recording with the Ninja Tune label since 1996. His first comic book was released with his debut full-length album, *Carpal Tunnel Syndrome*, in 2000.

He has toured internationally, both as a solo artist, and alongside such acts as Radiohead, Money Mark, Beastie Boys, Coldcut, Bullfrog, and Dan the Automator. He spoke to me from his home in Montreal, where he was preparing for his book tour.

It seems like you're on fire, creatively, making records, scratching, drawing, and writing books. Is that your handwriting and art on your web site, too?

[Laughs] Yeah, handwriting. I wouldn't exactly call it art. I don't call anything that I do art. It's handwritten, and I guess my DJing is hand cut. There's a people element in there somewhere.

Tell me about this book project you're doing.

This publisher, ECW, approached me to do a book because they read the comic book sleeve I did for the first album with Ninja Tune, *Carpal Tunnel Syndrome*.

When I was kid, I had story-book records, so I thought it would be cool to do something like that. So, for Ninja's first album, I did this little comic, and they liked it.

What was the comic about?

It was this lost DJ character. He's struggling to find his voice somewhere, playing at a place called Nuphonia, which was a club where people would go and have no fun. He would play for confused people, and the crowd got rather irritated and would start throwing objects, and he hurts his arm.

One of the tracks, "A Night at Nuphonia," is a narrative, and this was to make a little radio play happen.

Where do you come up with this stuff? What's your creative process like for coming up with stories and things?

I think that you can tell, it's all kind of the same to me. I might have a sketchbook on tour to doodle stuff. It could be track ideas, it could be some recording ideas, it could be some cartoon or comic ideas. I don't really see them as differences in format.

You can tell stories through sound, or you can tell stories through pictures.

The stuff on that first album, I wouldn't categorize as songs, necessarily; I think of most of it as little radio skits. Using the turntables, I have access to sound-effects records as a production tool and as a performance tool, and I just feel out the experience of it. Add the little dimensions in the back. It's a kind of sound designing, in a way. Something I want to do, one day, is do the Foley effects for a movie.

Do you like to have a story line when you're putting together tracks?

Sometimes, I might start with an idea of an actual narrative with characters, including protagonists, and I might go on to a more classic story.

Or it could just be sound.

I was really inspired by a *Muppets Show* album when I was young. They had a song where Gonzo was eating a rubber tire to "The Flight of the Bumblebee."

Gonzo was just munching on this tire throughout the whole song, rather quickly. It cracked me up as a kid, and still cracks me up. It was neat to envision this character eating a tire to this song. It put fun ideas to the picture.

What drew me to turntables in the first place was the fact that there was a lot of range. All your source material affects what comes out. You have almost unlimited source material to scratch from, and then your scratching is going to become just as diverse, in a way.

Tell me about your musical background.

I come from a classical music background.

I started studying classical piano when I was four, and I played for about 10 years.

When I got to that teenage period, when I wanted something to do, I genuinely got interested in it.

Instead of just doing it for your parents.

Right, as opposed to something that was forced on me. When I first heard scratching, from playing piano, I knew it was being performed, not something that was programmed on a computer. I heard a Mr. Mix solo from the 2 Live Crew in Miami when I was 12 or 13. I thought, "This is so different from my musical experience, right now, with classical music."

Classical piano seemed quite stringent. There wasn't much room for personal expression. At the time, I was too young to know how you could play a classical piece and express yourself.

When I heard scratching, it seemed like there were less rules than classical music, but there were fundamentals to it.

Where did you hear this Mr. Mix solo?

It was playing at a store, and I was there, hanging out with my sister who was buying a Cure record or something. It was playing over the speakers. I could tell that someone was making these noises, and I didn't know how. I could tell it was being performed, and the whole idea that someone practiced to proficiently make these noises somehow. Immediately, my ears perked up, and I thought, "What's going on here?"

I went up to the clerk, and he said, "Oh, it's a Mr. Mix solo."

And I asked, "Can I buy any of this stuff here?"

He sold me a couple records. I was in Vancouver, and I didn't really know the DJ scene. I would look at the cover of the record, and see this box with switches on it. I would go to Radio Shack with my paper route money and say, "I think I need one of these boxes!"

So, once you figured out there were turntables and mixers in the world, what happened then? What'd your parents think?

They didn't really know what was happening. They just knew I was spending a lot of time in my room.

When did you first get a pair of decks?

I had a hi-fi. Well, I didn't really have a hi-fi, it was my sister's. I would sneak in her room when she wasn't there. It was one of those systems that had a radio, tape deck, and turntable on the top, and it was all connected, and it was terrible. Everything was made out of plastic. I tried scratching on it, but it would jump the needle all the time.

So, I figured out a way to do it. A friend of mine was working at this burger joint, and he gave me this wax paper that they wrapped burgers in. I made this slip mat, and a lot of it was trial by fire.

My first set-up was my sister's stereo with these burger wrappers cut into 12-inch circles.

Fig. 5.2.2. Kid Koala demonstrates his "flexi-disc" technique.

I used that for two years. I mainly scratched one of two records. They were Flexi-discs that came in magazines, and I found that they were the only records that didn't skip this needle. One came with a *Time/Life* record, the other came with *Modern Guitar* or something like that. But I would scratch—mostly, the guy talking.

Was there any way to control your attacks like you would with a fader or switch?

This knob that engaged the phono, and right next to phono was AM radio. I would tune the AM radio to this silent static—somewhere really quiet—and that was my cut-off switch. So, I learned my basics doing that.

How did you learn scratching, did you have videos?

No. Vancouver in 1978 didn't have any DJ shops. And I wasn't really old enough to get into Hip-hop concerts at the time. I was mostly listening to tapes.

In the meantime, I was delivering papers and trying to save money to buy a set-up. When I first could afford to, I bought one turntable and one mixer.

What'd you get?

It was a Gemini thing, a 1600, or something like that. Had a little pyramid on it, and this fake wood-paneled mixer! It didn't even have crossfader on it. I would just kind of use the switch, at the time.

I bet you were getting good with the switch by then!

Yeah! [Laughs] So, that's what I started with. When I was in grade 10, a friend of mine got these two 1200s and Numark mixer. I was like, "Oh man!" And he would barely practice. I would always go to his house and jam and work on stuff. That was pretty much most of my early teenage years.

When did you get a decent set-up of your own?

Not until my first year of college. I was mostly using my friend's all through high school. At some point, he just left them at my house. Then we graduated high school, and I went to University, and found myself back at the Gemini, back to the first turntable.

In high school, did you play any mobile DJ gigs?

Yeah, like, parties for sweet sixteens, which was cool because we were only thirteen. It was tedious, at the same time, having to take requests for records that you didn't really have yet because you didn't have the money. We would spend most of it on stuff that we wanted to hear. Unfortunately, we couldn't find the context to play them all.

Would you scratch at your mobile shows?

Sometimes. You know, it's not really about the DJ at a sweet-sixteen party! [Laughs] We'd just try to throw it in there, for fun.

Did you keep up your piano playing as well?

No, I pretty much abandoned it. Not until this year, did I pick it up again.

Now, I use keyboards to make beds, like chord beds, so I can scratch over them.

And the soundtrack that comes with the book, it's piano with turntable over it.

We wanted to stay in line with the feel of the book, which is like a silent movie because there's not much text. You can do all your own dialog, assigned to the picture. So when it came time to do the score for it, I felt it should be kind of quiet and maybe be piano with counterpoint turntable stuff.

Fig. 5.2.3. Kid Koala simultaneously operating the 1200s and a Wurlitzer electric piano.

You majored in elementary education at McGill, is that right?

Yeah, it was a pretty good way to inherit records.

I hadn't thought of that! Schools would give you their record collections?

They gave me at least 200 records during my practicums!

A lot of the records looked like they were rolled over with Tonka trucks. [Laughs] Still, I'd find parts …

Did you do any music at McGill?

No, I didn't. Elementary education is a four-year program there, and it's pretty much a lot of talking through psychology. The last year, you're all sent off to the schools.

Did you ever teach?

I graduated from McGill, and immediately, my mom would call me every day, and ask, "Did you get on the substitution list somewhere?" And I'm like, "Mom, I just graduated."

In the meantime, all through college, I was DJing, and I started playing with this band, Bullfrog.

Did you hook up with the Bullfrog guys at McGill?

No, actually they were all older than me. At the time, I was 19, they were all 25, 26. They were out of school, playing on the circuit, and started this band.

There's a nightclub called Metropolis—sort of a big concert hall that had three different rooms. I was hired to do the Savoy Lounge upstairs every week with another DJ, and we had this thing on three or four tables.

One guy would play records and one guy would scratch, and we would take turns doing that, and every week, the promoter would hire new musicians to come play with us.

One day, he invited Mark Roberts, who is the guitar player for Bullfrog, and musically, it clicked pretty quickly. He told me about this rhythm section he was working with on Tuesdays. I had never considered playing with a band in a live context, I was heavily into the DJ world, at that point in my life. So I was making excuses not to go.

They'd say, "You should come down and check us out!"

And I'm like, "Yeah, but it's cheap movie night!" [Laughs]

Finally I went down and checked them out. They were really, really tight, like no band I had heard live. They were rooted and focused on funk elements, which was where a lot of the Hip-hop records I listened to came from. So, I thought, "Yeah, I can follow this. I can probably find something to do here."

We started playing together, and that was it. At first, it was kind of funny. They'd be like, "Okay, give it to the DJ!" and they'd take it down to just the hi-hat hitting eighth notes, and I'd be up there cueing like, "You bastard!"

We all had to get used to each other. They didn't really know what was possible. A lot of stuff I was throwing at them, they were like, "Whoa, this is kind of a weird groove . . . Kind of square."

In the meantime, they're throwing all these shuffles at me, and I'm thinking, "Whoa, what kind of stuff is this?" I had to learn all these new patterns on the decks.

Musically, that was a very big learning experience for me. We played every Tuesday at the Voltaire nightclub for two years, and tried to have some new things thrown into the mix every week. And we were getting booked in other cities, so we would drive out on the weekends.

It was grounding. I think that sort of cracked my head open to a lot of how to blend. At first, I felt like I was just scratching over what they were doing. But as we played more together, we started actually writing together, and finding ways to integrate me into the song, as opposed to, "Take it here in the bridge, and now back to the music."

Instead of being the turntable novelty . . .

Yeah. I was resenting that, at first. When we started writing together, I was more integrated into the band.

How do you think your classical music background affected how you integrated with the band?

I don't know if I can put my finger on it. I think being raised listening to music that way and to the mechanics of it . . . I mean, they'd use terms that were over my head when talking about music. When I first started playing with them, it was like, what are you talking about?

What kinds of things?

They would talk about paradiddles and stuff to drummers, and I was like, "I don't know what that is." And even terms like "bridge!" [Laughs]

I remember, when I was learning things as a kid for recitals, it was, "Learn every note on the paper exactly like the dude wrote it. Don't think!"

With the band, I learned more about certain styles of song structure. Not that I've used it in my own music, but when playing with a band, there's definitely a higher order of what's good for the song. We're not just here to show off.

Did you ever do DJ battles?

I did one competition, one year, when Bullfrog was just getting started. I did the DMC local, I didn't place in Montreal, and I didn't place in the Canadians.

We did this other one, I forget what it was called, but this radio station put it on in Toronto. We ran out of gas and had to rent a U-Haul! It was terrible, we were the two DJs from Montreal, the other DJs were all from Toronto. We were driving, and my man's car broke down! He was the DJ that I was playing with every week.

So, we're in Kingston, which is sort half-way from here to Toronto, and no one would rent us a car, so we had to get a U-Haul!

We had no steady income, not enough to pay rent, but we had some record buying money, so we decided, "We should do it, we should go get this U-Haul."

So, we get this BIG U-Haul, and all we had in the back were two turntables and some mixers. We were driving, and the U-Haul was just guzzling gas, so we get to Toronto, we're listening to the radio, and we hear, "Yeah, we're still waiting for Koala and D. We think they're driving up from Montreal, and if they're not here soon, we're just going to skip them."

I'm like, "Drive! Drive!"

And then we hit all this rush hour traffic. My man Danny is super composed. For the most part, he's not a man of many words. He's just very quiet and practices scratching all the time. It was terrible, and all of a sudden, we're two blocks away in this U-Haul and, and he says, "NOOOOO!"

And he stands up, sort of doing an arm press on the steering wheel. I'm like, "What's happening?"

And he yells, "We're out of gas!"

Meanwhile the radio is still on tuned to the broadcast of this competition, three of the guys have already performed, and I'm like, "WHAT?! NOOOO!"

And we ran two blocks, and I bought one of those red canteens and filled it up with gas, I RAN back, we drive, we make it right to the venue at this community center.

We open the latch and grab our turntables, we run, like totally stressed out. Hit the stage, we put the decks on the table, and the MC is there, and he's like, "Who are you?" and I'm like, "I'm Koala, dude, from Montreal!" and he's like, "You guys are disqualified!" [Laughs]

Oh no!

[Laughs] YEAH! They're like, "You know you have to be here at whenever," and I'm like, "Dude!" It was awesome.

So, you didn't get to play?

Well, we did get to play, as sort of a showcase, but we weren't allowed to compete. We would've lost that day, for the record, 'cause the guys that took it were dope.

So, that's what happened. Short-lived battle career. Part of me was asking, "Do I want to keep doing this or not?"

You're obviously connecting with an audience, and a good portion of that audience is DJs.

I can talk to DJs and relate to them probably better than a lot of other humans.

The thing with DJs is all that time in isolation. People think that you're like Trekkies, or something, because you can talk about DJing for hours.

Fig. 5.2.4. Koala's vintage Victrola, using a Shure SM-57 mic rather than a horn, is a faithful companion on the road.

It's kind of a solitude-friendly craft. Most of the DJs I meet, despite what their personas are in a battle, are really mellow.

How did touring with Radiohead come about?

It was weird, when Radiohead asked me to do that! I was listening to *OK Computer* straight from when it came out for probably about four years. It was really such a wonderful record.

And when they asked me to tour, my first thought was, "Oh yeah! That'll be great!" My second thought was, "Oh no! What am I going to do?"

I'm a Radiohead fan, myself, and I know what vibe I'd probably want to be in, if I was at their concert. What I was known for was weird carnival music.

You know, I tried some new routines that I think maybe worked, and I was surprised that the audience was very open-minded.

How would you describe your Radiohead show?

I had a 20-minute warm-up set. We were playing places like the Hollywood Bowl and Madison Square Garden. I knew it wasn't going to turn into a B-boy B-girl breaker party. So, I opted away from a club set, and went for a more … I don't know what you would call it.

I did one of my jazz trumpet routines.

I did something with some Radiohead samples, which, in some ways, is kind of cheap because everyone's going to cheer for the Radiohead samples. But I was alone up there!

Believe it or not, even if I wasn't at that show, I was integrating Radiohead into my club sets. If I was out touring on my own as a DJ, I would use some of their records.

Do you think that working Radiohead tracks into your set had anything to do with them asking you to play?

I don't know, because the first time they saw me, I didn't do any. But it was weird, I was such a huge fan of theirs, but I didn't really see any obvious links musically between what we were doing, in terms of recording.

But I realized, when I started talking to the [Radiohead] guys while we were on tour, they can quote lyrics from Public Enemy's second album, which was the blueprint for a lot of DJs. I thought, "Wow, you guys listen to that music?"

You've also toured with the Beastie Boys?

Yeah, that was with Money Mark [Beastie Boys keyboardist].

Tell me about Money Mark.

I think that man's just consistently five years ahead of his time. That was back in 1998, and I was working on *Carpal Tunnel* at the time, and having a horrible time.

In what way?

I knew that as a first record, I didn't want to do stuff that I had done before. I wanted to try going to other areas and seeing where I could take it. A lot of uncharted territory for me, and obviously, the anxiety of it … But I needed to do it, as a learning experience, as well.

It was kind of this weird, mad-scientist feeling.

Money Mark called me in the middle of that and was like, "Hey, do you want to go on tour for *Push the Button*?"
This was his second album that came out, and I was always just a really big fan of his. I was like, "Wow, that'd be awesome!"
So, he flew me out to LA and called me up to join his band on tour. We did a couple short tours, just as the Money Mark band. After that, the Beasties released *Hello Nasty*, and he was on that tour, and had the opening spot. We were rolling with them for that summer.
It was fun, because after those shows, Mark would do his own little shows, and Mixmaster Mike would come out, and me and him would do a little opening set for Mark at this bar, after. It was a great tour. We had a lot of fun.

What's up with Dailey City? Have you ever wondered about that?

Maybe some semi truck full of turntables flipped over in that neighborhood, and everybody had a pair, and everybody got really good, really fast! It is a Mecca for DJs, over there. I always felt a little intimidated there, like, "Yeah … okay." Their grandfathers can scratch better than me.

How do people respond to Bullfrog at jazz festivals?

They seemed to enjoy it. A lot of the audiences haven't seen anything like us before. Bullfrog is a party band.

Our first few gigs were opening for Maceo Parker, and we learned a lot from him. I'd heard him and his band on record, but when I saw them live, I was just taking notes.

He toured with James Brown for how many years?

What do you feel like you learned from them?

Everything! Having so much fun on stage, this exuberant, sort of contagious energy in the room. It was amazing.

Fig. 5.2.5. Koala adjusts his mixer during sound check.

In your solo work, you seem to be constantly doing things differently from other people.

I don't know if I'm trying to do something differently from other people, or differently for myself. I'd get really bored if it was like, "Here are the same five songs you have to play for the rest of your life!" Of course, I say that now, but you'll probably catch me … [Laughs]

Isn't that just a tragedy of being a DJ in the first place? You probably can't sing, you're probably too short to play basketball, everything's been said, all the music's been played, and somehow, you're supposed to put something out there with your personality in it.

I was speaking to this selector DJ once, and he was talking to me about the art of selection. He was kind of like Mr. Miyagi, because at the time, I was trying to grow and see a bit wider picture. The whole scratch revolution was happening, and people were very into it and closed to other things.

He was talking about what he does and how long he spends just listening and learning his records. Learning segues, learning transitions, learning how to time things, like blends. He was talking about holding the mixes in line, which is something that I'm not too good at or interested in because of my short attention span.

But at the same time, I was really inspired by the honesty, and I realized that it was the same thing in scratching, or house DJing. Even if it's playing other people's records, he said, the selection and order you put them in will affect the mood in the room and will affect whether people are enjoying whatever context you happen to be playing in.

He was talking to me a lot about transitions, and he had it down to a science. I realized that this guy spends just as much time learning all the nuances of his records as I do working on this one scratch.

That's when I started digging for weird things on records. It's the fisherman in me.

You drop your lure, you drop your needle, and you just let it sit there, and sometimes you catch stuff, and sometimes you don't. I think that whole part of it is just fun, you know what I mean? I don't do it because I have to find something different, I do it because I think it's fun. And when I hear something interesting, I think, "Man, it'd be cool to freak that out," or, "Man, it'd be cool to have two copies of that," and that's how it goes.

It really starts from the records. And I can't really prove it, but I think, in some ways, the types of things you select and put together, somewhere in there—even though all the source material is someone else's music, or someone else's samples, or someone else's thoughts being spoken, just the way that you cut it up or edit it—I think that somewhere in there, you can get your personality.

It's a hard thing for people to understand.

My mom still doesn't get it. She asks me, "So, when are you going to write your own songs?"

I think some people can't see past the collage aspect of it, and it has to be "completely from scratch." And how far do you go for that? Where did the paper and the pen and ink come from?

At this point, I'd actually like to be part of the confusion.

Some of what I do is generated completely from scratch, and some of the records that we press now are sounds that don't exist. They're actually just tools that we invent for ourselves.

I think one of the main rules in music is just do something fresh. I apply that philosophy to anything that means anything to me. And with scratch music, that's the whole spirit of it, you know?

6

The Rise of the DJ As Musician

The notion of the DJ as musician strikes many as an oxymoron.

When Kool Herc began to repeat sections of a record with two copies, he started down a path of changing the music rather than just playing it back. Yet, that makes one an editor, not a musician. As Flash added embellishments of kick drums and other sounds, the turntable stopped being just a playback medium, but did it become a musical instrument?

As a new idea called a "mash-up" gained momentum, DJs began creating new pieces of music by taking parts of two or more records and combining them in creative ways. Yet, even that makes one an arranger, and not necessarily a musician. Perhaps the most closely related analogy in the visual realm, the collage, may provide some illumination. When Matisse began creating collages, some questioned the art form; few today would argue whether or not Matisse was an artist, or whether a collage should considered art.

At what point does a DJ become a musician? At what point does the turntable become an instrument? What does a DJ-turned musician bring to a performance that a guitarist or drummer does not?

The basic elements of music—pitch, rhythm, harmony, and timber—are all under the control of the master DJ. What each DJ decides to do with these elements can be vastly different.

I've heard some people argue that since DJs are playing back music written and performed by other musicians, they themselves cannot be considered musicians. This argument breaks down when one considers that since the 1960s, musicians have been involved in using recordings of other musicians to make new music. This era was ushered in by the invention of the Mellotron in 1965. The Mellotron was a keyboard instrument that played back tape recordings of strings, recorders (the flute kind, not the tape kind), choirs, and even entire rhythm sections playing various grooves. You know the odd recorder parts on the beginning of "Strawberry Fields" by the Beatles? That's being played on a Mellotron. John Paul Jones (Led Zeppelin's bassist and keyboardist) used the Mellotron in concert to reproduce the recorder parts in the beginning to Stairway to Heaven, and the Moody Blues used the instrument's pre-recorded sounds extensively.

This trend became a title wave with the introduction of digital sampling and MIDI (Musical Instrument Digital Interface) in the 1980s. Sampling allowed digital recordings of instruments to be stored and played back with much more flexibility, reliability, and stability than the Mellotron, and MIDI allowed these recordings to be triggered by keyboards, drum pads, guitar controllers, wind controllers, and all manner of devices used by musicians to play back and manipulate the recordings of other musicians.

The president of a Japanese DJ equipment manufacturing company once put forth to the president of Berklee College of Music that the difference between a DJ's and a MIDI musician's various manipulations of pre-recorded material was that the DJ was working on the "musical phrase" level, while the MIDI musician was working on the "musical note" level. While there may be an illuminating piece to this analogy, it is an oversimplification.

Anyone watching QBert drum-scratch will realize that he is isolating individual kick drum, high-hat, and snare attacks. DJ Radar is playing melodies in his turntable concerto, and seldom playing back an entire phrase from a record.

For many, the concept of using the turntable as a musical instrument seems farfetched. It's difficult to accept that a device invented for one specific purpose, in this case *carefully* playing *fragile* records, has been reborn in such a radical fashion.

It may help to consider that, if you go back far enough, every instrument in the symphony orchestra has origins as something else. Musicologists trace the birth of instruments using vibrating strings back to the bow and arrow. Clubs, hollow logs, wine skins, gourds, and blacksmith's tools eventually gave us the modern percussion section. Blades of grass and hollow reeds evolved into woodwinds, and people blew into the horns of bovid animals in order to signal danger long before some zealots with too much time on their hands practiced blowing long enough to play melodies.

So when does an object become a musical instrument? As with the examples above, the key is the intention and resolve of the person operating it.

"I hate to view this technically, but to me it's an instrument if it's used like an instrument," reasons Jim Tremayne, Editor of *DJ Times Magazine*. He's not alone.

"A musical instrument only becomes a musical instrument when a person uses it for that purpose," says Jazz pioneer Herbie Hancock. "A turntable becomes a musical instrument when the person behind it is creating some musical ideas."

Perhaps the first instance of this is John Cage's piece, *Imaginary Landscape No. 1*, written in 1939. Cage composed *Landscape* for muted piano, Chinese cymbals, and two turntablists playing Victor Frequency Records. The composition takes advantage of the turntable's ability to alter the pitch of these tones by varying the speed of playback.

Thirty years later, many factors conspired to accelerate the development of the turntable as a musical instrument. The popularization of Hip-hop brought the newly developed scratching, cutting, and back-spinning techniques of Grandmaster Flash onto the world stage.

MTV launched soon after, ushering in the era of music video. Herbie Hancock's hit record *Rockit* featured Grandmixer D.ST's scratching front and center, and Jam Master Jay's skills behind the decks with Run DMC, along with Jazzy Jeff's transformers and chirps behind the Fresh Prince gave the whole world a first-hand look at what a Hip-hop DJ could do.

Thousands of impressionable kids took notice, and soon young virtuosos who called themselves turntablists were overrunning DJ competitions like the New Music Seminar and the UK's Disco Mix Club (DMC).

The proliferation of inexpensive home video tape machines made it possible for those practicing in their bedrooms to study the world champion's winning techniques within a few months of the competition.

Turntablists were popping up in other forms of music as well. Innovative artist Christian Marclay combined Hip-hop turntable techniques with contemporary electro-acoustic music and performance art esthetics to create pieces employing the use of up to 99 turntables at one time.

In the new millennium, most nu-metal bands on the Ozzfest tour employ turntablists, and pop icons as accessible as Sugar Ray, Madonna, Kid Rock, and Janet Jackson or as eclectic as Portishead, Incubus, Beck, and Moby feature turntables as a part of their band.

In addition to Herbie Hancock, jazz artists like Christian McBride, Liquid Soul and Medeski, Martin and Wood have made it likely that turntablists will be taking the stage at most major Jazz festivals for the foreseeable future.

The musicians on the following pages are pushing this movement forward.

6.1

Herbie Hancock: The Inspiration of the DJ

If you are wondering why a book about DJs features an interview with the jazz pianist who played in arguably the best quintet ever fielded by the immortal Miles Davis, consider this unlikely fact: Herbie Hancock has had more influence on turntablism than all but a handful of DJs. But then, Herbie Hancock has had more influence on popular culture than any other jazz pianist, living or departed (Figure 6.1.1).

Hancock introduced synthesis to jazz, and jazz-fusion to the world on Miles Davis' pivotal records *In a Silent Way* and *Bitches Brew*. *Headhunters*, Herbie's 1973 release including the hit "Chameleon," became jazz's first platinum album. The 11 albums Hancock placed in the pop charts in the 1970s have provided Hip-hop artists in the following decades a constant source of samples and inspiration.

In 1983, Hancock partnered with producer Bill Laswell to create the platinum album *Future Shock*. The smash hit single and music video "Rockit" featured Grandmixer D.ST (now DXT) scratching up a storm, and sent a young Richard Quitevis scrambling for his parents' turntable, along with seemingly every other kid in South San Francisco and Daily City (if not the world). "Rockit" won a Grammy, and helped usher in the music-video era by lending credibility to this new art form, racking up five MTV awards in the process.

In 2002, Hancock and Laswell teamed up again for *Future 2 Future*, featuring appearances by Grandmixer DXT and the X-ecutioner's Rob Swift at the decks. Daily City's DJ Disk has been accompanying Hancock live, as Herbie continues to feature turntablists on equal footing with other instrumentalists.

An articulate philosopher and visionary innovator, Herbie Hancock, has helped open people's minds to cultural and musical diversity for decades. I recently spoke to him about the evolution of the DJ, and his unique role in that evolution.

How would you explain your collaborations with turntablists to people who have trouble seeing the turntable as a musical instrument?

A turntable becomes a musical instrument when the person behind it is creating some musical ideas. Using the turntable for scratching is one example—then, it's a rhythmic thing.

Another thing is, very often, turntablists play bits and phrases of tracks from other records, whether it's a spoken word record or whether it's a record from a TV show or a radio show. Or they might use something from a madrigal, or a choral, or music from some other country.

Fig. 6.1.1. Throughout the past five decades of his professional career, Herbie Hancock has personally ushered in an astonishing number of new movements in music, art, and culture.

A musical instrument only becomes a musical instrument when a person uses it for that purpose. A piano sitting in somebody's living room is a piece of furniture because it's a piece of furniture.

If you can take a wicker basket, put some beans in it, use it as a shaker, and call that a Brazilian musical instrument, then you should call a turntable a musical instrument. Or anything else. A doorknob could be a musical instrument.

What was your first exposure to scratch DJs, and what did you think?

The first time I heard scratching was on a record called *Buffalo Gals* by Malcolm McClaren. The week after that, we were doing "Rockit."

How did you put "Rockit" and "Future Shock" together? Who came up with what?

Well actually, Bill Laswell at the time had a partner, Michael Beinhorn. They had a group called Material, and I didn't know anything about them. But a guy who was working for me at the time, Tony Milan, one of his jobs was to kind of snoop around and see what's going on in the music scene. See who the new people are, what new developments are happening, and just kind of get a perspective on a lot of things having to do with art and culture. Tony knew about Material, and he thought that if I worked with Bill and Michael that it would be a match made in heaven. Here are two very eclectic guys working together with me, also very eclectic. I brought things to the table that they couldn't bring to the table, experience-wise. Yet, they were in touch with some underground scenes that I didn't know anything about.

Bill agreed on spec to produce one tune, and I didn't have to use them if didn't like them.

The funny thing was, the week before they were to come over and bring me whatever it is they were working on, I happened to hear this record *Buffalo Gals*, at the suggestion of a young friend of mine. The kid's name is Chill Factor, but his real name is Chrishner. He's the son of the bass player Walter Booker, who worked with Cannonball and Nat Adderly for many years.

I started asking him, "Hey Chill, what's some of the new stuff that's happening?" He'd play me some new records. He brought me this record of Malcolm McClaren, and I heard scratching for the first time. And I said, "Hey, I like that! I'm gonna use that on my next record."

I was prepared to tell Bill Laswell and Michael Beinhorn about scratching, but before I got a chance to do that, they started playing the tracks that they had, and the first track that they played had scratching on it.

And I said "Yeah! That's what I wanted to do!" So, we started to work on it, and that became "Rockit." And then during the middle of that process, I said to Bill, "Look, let's do the whole record like this. This is what I wanna do."

Fast-forward 20 years, and tell me about how *Future 2 Future* came about—what you set out to accomplish.

Future 2 Future came about when Bill Laswell and I were having a conversation, primarily about where we are in life. The issues that are important to us, just trying to get a new perspective of where we were coming from. We hadn't worked together in over 10 years, maybe 12 or 13 years, and that period of time is long enough for people to make some evolutionary changes in their perspectives. And we saw eye-to-eye on things we felt were important.

So, during that conversation, he asked if I knew about the electronic scene, and I said not really, I know it exists, but I don't know very much about it.

The next question he asked me was did I know that some records I made in the past were a big inspiration for a lot of people that were creating this kind of new scene?

Since the scene kind of grew out of Hip-hop, I assumed that he was talking about "Rockit." But he said no, not "Rockit," but rather *Sextant* and *Crossings*.

I said, what? Those are very far out and complex avant-garde jazz records. I didn't know anything about it. He said it might be interesting to see what the result would be if I collaborated with some artists that he would hand-pick, and see what would come about, because they were influenced by records I made when I was their age.

Like a full circle.

Exactly. If their output was affected by the stuff I did, how would I react to their output?

Did you listen to records they had made before working together?

No, what Bill wanted to do was use a completely different approach. For me *not* to listen to what is being done in the new electronic era. I guess he was thinking that I already have an input in that direction, through people who have been influenced by me. A piece of me is there already.

What he wanted to do was make the approach to this record different from what we did with *Headhunters*, in that he wanted to make this a very spontaneous record. In the studio, he would play elements that he and other people put together for me. The first time I was listening to them, I was actually in front of a keyboard and the record light was on. He wanted my immediate gut-level response to what they were doing. That's how we put the record together.

Could you talk a little about the track "Rob Swift" on *Future 2 Future*, and how that came together?

All those people that I collaborated with, Bill actually got them. I actually didn't meet them until after we did the record. They put their stuff on first, and then I added things, took things away. The editing process was not just something that was done in post-production, but was part of the creative process of making the music.

How did you come together with DJ Disk?

He was also at the suggestion of Bill Laswell. Bill thought it might be interesting for him to work with me, interesting for both of us. DJ Disk, his whole family is musical. They're all musicians—his mother, his father. I've noticed that he has a very unusual and advanced rhythmic perspective. And also, in any musical environment, he's able to hear something that he's able to put in, in order to embellish that environment.

Do you see similarities in jazz and Hip-hop cultures in terms of the way they've developed over the years?

I actually don't know that much about Hip-hop. I mean, not like people would expect, considering the effect of "Rockit."

"Rockit" had a big effect on dance music in general.

Yeah, it kinda opened a big doorway. As far as comparing the development of jazz to the development of Hip-hop, it's hard for me to compare them. Jazz had such a long development, originally growing out of the blues, which grew out of slavery and the black church, and the evolution of jazz went through different periods, too. The 1920s when it was Dixieland, and the swing era, and the bebop era, and post-bebop and avant-garde, those took a while to develop.

On the Hip-hop scene, the idea as far as turntablists are concerned, is to become a virtuoso of that instrument and that could be compared in its own way to Charlie Parker becoming a virtuoso on the alto saxophone.

In previous periods in jazz, being a virtuoso was not a generally accepted goal. For some, it was. People like Don Bias and Lucky Thompson were virtuosos. You know, other guys, they played well, but as far as being technically virtuosos … even Miles Davis didn't profess or try to be a technical virtuoso. Dizzy Gillespie, maybe more so than Miles.

But on the Hip-hop scene, I can't tell you much about the other aspects of it—like rap, for example—because on the records I've done, there's only been one tune that was a rap. One of the guys who wrote it was Chill Factor, who first played scratching for me, and he wrote this rhyme in reference to me, which is something that exists in the Hip-hop scene. They pay homage to each other and themselves, you know. That's different from jazz. In jazz, I can't think of anyone who's paid homage to themselves.

Not so much overt bragging in jazz.

Right, right. One thing that I noticed from some of the best turntablists, for example, is they open themselves up to using records from a vast number of sources—a vast number genres and countries, as their source. In that sense, it's eclectic, and jazz too, in many senses is eclectic.

Have you seen a difference in the level of skill among turntablists in the 20 years or so since "Rockit?"

Oh yeah, that's changed considerably, I mean, it's broadened. DJ Disk uses one turntable. He does it all with one. That's his trademark.

But even Grandmixer DXT, when we're doing "Rockit," the live show, he figured out that it would be kind of interesting if he used effects on his turntables, 'cause no one was doin' it, then.

So, he bought some pedals and devices to use for echo and repeat and those things, back then. Since then, other turntablists have jumped on that bandwagon, and I think that's become more of a standard approach, expanding the possibilities of sound from that instrument.

DXT's not only a turntablist; he plays drums, he plays keyboards, he plays bass, he plays a lot of different instruments, so he comes from the heart and the perspective of being a musician in the traditional sense. And he's a great conceptualist. He's a smart guy.

Why do you play with turntablists? What does it offer your music specifically?

For me, the best thing about it is that I get to hear from one person, all kinds of different techniques to develop rhythm. Elements that I would describe as percussive elements, like a percussion instrument. Scratching is one of them. Depending on what they scratch from, the sound it makes can vary.

A single person can have widely varied sounds that they use for scratching and other kinds of percussive effects.

In that sense, you could compare turntablists to percussionists. That's one of the things.

But the other thing that turntablist can do is that they can function like another keyboard player and have different environmental sounds that paint the background and the tone of the musical experience, and the shape of the musical experience.

6.2

Logic's Project: Evolving from DJ into Musician

Lee Jason Kibler, a star basketball player from the Bronx who was about to graduate from Harlem's noteworthy Rice High School, was facing a pivotal choice. Should he accept one of the basketball scholarships being offered to him or pursue music full time with the alternative rock band, Eye and I? Rock music was an inhospitable path for black musicians to begin with, and at the time, playing turntables in a rock band was literally unheard of.

Jason took a chance on a music career, and within months, his risky choice was vindicated. Eye and I was signed to Sony's 550 label, and then picked up by the William Morris Agency. The Black Rock Coalition took shape with the bands Eye and I and Vernon Reid's Living Color in the forefront, and Kibler spent his 18th year traveling across the globe performing with artists as diverse as Ice T and the Psychedelic Furs.

Born in the Bronx in 1972, Jason grew up in the pulsing heart of the Hip-hop revolution.

Watching and listening, as Kool Herc, Afrika Bambaataa, and Grandmaster Flash took this new art form from neighborhood block parties to the rest of the planet, young Jason became an early Hip-hop historian. Soon after he received his own set of 1200s on his 14th Christmas, Kibler became the neighborhood DJ, playing at friends' parties and school events.

Jump forward to the mid-1990s, and Jason (dubbed "Logic" by Eye and I singer DK Dyson) was again on the cutting edge, experimenting with jazz and other forms of improvizational music. After recording and touring with Graham Haynes (Verve) and Don Byron (Blue Note), Logic joined Vernon Reid in New York for a weekly improv night at CBGB and the Knitting Factory.

Medeski, Martin and Wood soon asked DJ Logic to spin a warm-up set to kick off their run of "Shack Party" gigs at the Knitting Factory. When they invited him to sit in, magic happened. The audience went crazy, and Logic became the unofficial fourth member of MMW for the rest of the Knitting Factory run, was invited to play on the *Combustication* album (Capitol/Blue Note), and went out with MMW for about a year of touring.

Logic has had an amazingly creative run since then. His association with Ropeadope Records (Atlantic) has seen him release three CDs as a bandleader (*The Anomaly*, *Project Logic*, and

Fig. 6.2.1. DJ Logic shares his insights at the Berklee College of Music in Boston.

Zen of Logic), and one album as half of the mythical Yohimbe Brothers, with Vernon Reid on guitar.

Logic seems to have found his calling playing with other musicians, live, in real time.

Logic's tours with Ben Harper and Jack Johnson in 2003, and John Mayer and Maroon 5 in 2004 saw Logic spinning a DJ set to warm up the crowd, then joining Johnson and Mayer on stage during their sets, adding fresh rhythmic interest to radio hits like Mayer's "Your Body Is A Wonderland." Logic has forged a solid connection with the jam band audience, with high-profile performances with the Grateful Dead's Rob Wasserman and Bob Weir, as well as the String Cheese Incident and Bela Fleck.

Logic is also one of the highest profile DJs in Jazz, having graced the cover of *Downbeat* magazine. The first DJ to play the Blue Note jazz club in Greenwich Village, Logic's musical collaborations include work with Christian McBride, John Scofield, and Teo Marcero.

His modesty is legendary, and when he's not touring or recording, you can find him living in the same neighborhood that he grew up in.

I first met Logic in New Hampshire at an outdoor festival celebrating the anniversary of the Red Hook Brewing Company. He had the crowd eating out of his hand during his solo set. After our interview, he returned to sit in with the band Soulive, and to my surprise, flew in many horn samples from *Turntablist's Toolkit*, a record I produced as a resource for turntablists looking to incorporate elements of jazz.

Logic later became the first jazz turntablist to do a master class at Berklee College of Music. Here is our conversation from that evening in New Hampshire.

Let's talk about that set that you did today. This is a rural New Hampshire, white, jam band kind of audience. How do you approach playing a gig like this, as opposed to if you were in a club?

There are a lot of people out there that listen to all types of music. You see the young kids, Hip-hop kids, skateboarders, and you see the older crowd as well, so I try to entertain everybody.

I try to do things spontaneously that the crowd wouldn't expect, just to watch their heads nod after they hear James Brown into some bebop tune.

Who are some of your favorite DJs?

DJ Hollywood, because he was one guy who used to do big party jams in front of thousands of people. Flash, Kool Herc, Afrika Bambaataa. To this day … Pete Rock, Kid Koala, DJ Krush, DJ Shadow …

What did you get from growing up and hearing Flash, Bambaataa, and Herc right there in your neighborhood?

The knowledge of how they were having a good time up there, and how they were controlling the vibe of the crowd with the records they played. I saw how they were being creative by playing an eclectic set, using different types of music to entertain the crowd. All I really wanted to do was the same thing: controlling the crowd and just spinning all types of grooves.

At that time, they really didn't have the beat machines; they were using the turntable as the beat machine. For the MC to rap over, they would just have two of the same records and find that funky break from either, like, a Kool and the Gang or a Manu Dibango record, and they'd go back and forth.

How did you learn to scratch?

Going to those Zulu Nation anniversary parties and seeing DJs scratch, and hearing other DJs on the radio like Marley Marl, Jam Master Jay, Red Alert, and Mr. Magic. After seeing and hearing those DJs, I was like, "I want to learn how to scratch."

Also, I had a friend in my building who had turntables. The stuff that I heard, I would just try to copy the same type of scratch. I would go down to my friend's place, and he would teach me how to scratch. Eventually, he got tired of me coming down all the time and using his turntables.

It was time for me to ask my mother for some turntables of my own, and when I got turntables for Christmas, I was psyched. From there, I was in the house every day just practicing on the turntables. Every Friday, I would go down to the record store and pick up the latest Hip-hop and breaks records. Then I would come back home and use my new records to practice mixing and blending, and then I'd practice scratching.

I had a friend who lived next door to me who played drums, and he would go on tour all over the world. His name was Richie Harrison, and he used to play with Melly Mell and all the Hip-hop heads. Richie would tell me stories of the trips he'd been onto Europe with these guys, and who he's played with, and that got me excited.

One day, Richie asked me, "Hey, you want to come down and play with my rock band?"

And I was like, "Rock band?" At this time, in the 1980s, I was 15 and listening to Hip-hop.

He said it was his alternative rock band. I was curious, but kind of scratching my head. I said to myself, "What the heck. I'm 15, I'm still young and have nothing to lose." I just wanted to explore, so I decided to see what it was all about.

Had you ever seen a turntablist in a band before?

No, I hadn't seen a turntable in a rock band. This is like 1984 or 1985.

Talk about the turntablist's role in a band situation and how you come up with your ideas of what to do.

When I started playing with the band, I was just seeing how everybody was playing their role and playing their part. From the bass player, to the drummer, to the vocalist.

I look at the rest of the band like they're a third turntable. And then I think, "Hmm. What's missing here that I could fill in?"

I started kind of being a percussionist in a way. At that time I was doing a little bit of scratching, but I was also throwing in samples and sounds. And I saw myself as being a musician. I was part of the whole process when it was starting up, but to me I was just having fun and was into it.

That's how I met Vernon Reid of In Living Color. They got signed to Sony Epic, and we were the next black rock band to get signed to Sony behind them, around 1989 or 1990. Things just happened so fast, and I saw everything. I was like, "Wow … Hip-hop!" and then I went to this alternative rock thing.

Fig. 6.2.2. DJ Logic sitting in with the group Soulive at an outdoor festival.

What's next? I was just scratching my head. But we did very well.

During that time, around 1990, I also got turned onto jazz. At that time, Eye and I's bass player, Melvin Gibbs, was also thinking of a new idea: how a DJ would sound with the improvization of a jazz band. So, he used to do these improvizational jazz gigs at the old Knitting Factory. They wouldn't let a kid my age come in there, because it was an older crowd, so Melvin just had me come in a little early and got me in around the back door.

It was a whole different thing for me too, because everybody was playing out. Everybody was doing something different, but everybody was in sync, and everybody had a vibe.

The whole thing with jazz improvization is everybody feels each other. Everybody has to have a connection and feel comfortable with each other, because they're vibing off each other.

That's how things move smoothly.

So, I looked to the improvization of jazz band as a third turntable, and I was just trying to find my little element—what I could add between all this movement that was going on the stage. Everything was just happening so spontaneously, and to me, it felt like I had to be fast-thinking, you know? You're hearing certain things in certain cool places, thinking, "This would sound kind of cool with that." I was just adding a whole bunch of different sounds or doing certain rhythms that people wouldn't think of.

People would ask, "Wow, how'd he know that the rhythm was right there?"

The musicians might be playing one thing, and I'd be right there with them, but overlaying something they didn't expect. And we'd all be surprised how cool what we were doing together sounded.

That's how I hooked up with Medeski, Martin and Wood. Just playing around in the downtown scene with a lot of different jazz people. That's how we kind of clicked, actually. The whole *Combustication* thing came about right after their *Shack Man* album.

So, you'd already been playing with jazz groups.

Yeah, I was playing with jazz groups. I played with people I never thought I would play with, you know? And I have their records at home and I'm like, "Wow …"

Vernon Reid said that you're a musician first and a DJ second. Are you comfortable with that?

You could say that, because basically, before I start DJing, I'm listening to what's going on and then finding my "note," meaning that certain record or that certain sound or color that blends to what they're doing. It's just like if a guitarist is tuning his guitar before he starts to play while the band is in the groove already, and he's just tuning and getting ready before you even hear him. I'm DJing. I'm in the headphones and cuing up the record, just like tuning, and I'm finding the right color, and then I just bring it out, unexpectedly.

Fig. 6.2.3. Logic uses the turntable to emulate a variety of instruments when playing with a band, employing a variety of resources, including horn samples from *Turntablist's Toolkit*.

People will be like, "Wow, where's that coming from. Who's doing that? Cool."

It seems like you really listen to the rest of the band.

I think that's the most important thing, for all musicians. Everybody listens before they play, so it's the same thing with the DJ and your role as a musician: to listen to what's going on around you, to where things are coming from and where they are going.

That's the connection: not being afraid. I never went to music school, but listening to all these different instruments and the keys that they're in, when I grab a record, for some reason, I can just pick that up.

Like with those horn records you made. Boom, I'm in there and people are looking for the trumpet player or looking for the sax player. They're like, "Who's doin' that? Oh, that's the DJ!"

And that's cool, ya' know? It makes me happy just seeing people smiling, and going "Ahhhh."

When you're working on arrangements with other musicians, what vocabulary do you use? Are you talking about 16th-note triplets, for example, or are you more comfortable playing stuff for each other?

It's a variety of different things. I talk to them by using the turntable. I'll show them certain things, and they'll say, "Okay, you want it in that rhythm or you wanna take it there?"

Hearing them play, I'll be like, "That's it. Let's record that. Let's do that."

I might just be hearing a certain rhythm, and I'll talk to the drummer, and be like, "Yeah, give me a Hip-hop groove like this, or do the beat box, or quarter notes, or 16 trip. That type of vibe."

With the bass player or horn player … [sings in demonstration] … and they will do something in that range that they feel comfortable with, and it'll go from there. I really don't throw too many ideas at them because I want them to feel comfortable. Usually, they'll be like, "Well, let me play something and see if you like it."

We'll just take it or just jam—doing some improv playing, and then go back to those tracks, and start editing. Just cutting it up, and making it into tunes and stuff.

Having done this now for 20 years or so, how would you say the DJ scene has evolved over the last two decades?

[Chuckles]

It's evolved a lot! It's amazing, because there used to be one record store I would go to in Manhattan. Now I go to any music store and they have a section just for DJs! I just started seeing that in 2000. Everywhere—all of the Guitar Centers, all over. Every state I've been to and every music store I've gone into, I've seen a whole DJ section, and I'd never seen that when I was growing up, because it was a hobby! Something fun, for DJs who were mixing at that time.

Nobody expected it to get as big as it is. They expected rave parties, but not stadium rave parties!

You go out to Europe and other states, and you see all these DJs doing these rave concerts like rock concerts. Everybody is grasping it so fast!

Everyone wants to be creative—wants to learn how to draw. It's like drawing your first thing, and when you just get so good at it, and you continue and evolve into a vibe, and it's just beautiful.

Let's move from live into the studio. What's your motivation, and what's the process that you go through to create tracks?

At the moment, just thinking about something that might be interesting. Something I could imagine me spinning or other people spinning and other people just vibing to and enjoying. That's how I look at it. I see a sketch, and I'm just trying to fill in the colors.

Do you record to ProTools or what do you use technology wise?

Yeah, I record to ProTools and use Logic.

Do you enjoy remixing?

I love remixes. I enjoy doing remixes—being able to create, and changing a groove around.

I'll try to figure out where I'm listening to the groove, and how I'm feeling the groove.

It's just great doing remixes and manipulating and changing sounds, and stuff. Doing something that the original artist didn't think could happen. And once they get it back, they're like, "Wow, I didn't know it would be like that!"

I did remixes for Medeski, Martin and Wood for the *Bubblehouse EP* album. The song I remixed is called "Dracula." Then I did another one for them, "Start and Stop." After that, I did a remix for Warren Haynes and Government Mule, which featured Chris Wood and John Scofield.

I did a remix of this Japanese group, Kankawa122, and Rick Holmstrom, who's a great guitarist. Also something I played today was a remix of "Herb Man" for Olu Dara. That was special because I'm a big fan of Olu Dara, and I have a lot of records that he played on. Just meeting him and feeling his vibe, he was an awesome guy.

When you do a remix, how do you like to work? Do you like to acquaint yourself with the original mix?

Yeah, I like to listen to the tracks already mixed, first, a couple of times. Then I ask to get the tracks broken down—different individual tracks, so I can have something to play with and change around. I just take certain elements that I think would be cool to use or loop, and I'll just start messing around with the loops and come up with a beat. I try to get the track lined up with a click—with the groove.

It seems like part of the evolution of the DJ is into remixing and into producing.

Yeah, you know just making that groove a danceable and interesting groove for whatever it is—lounge, house, Hip-hop, or jazz in general. Some funk.

What's the discipline part of DJing for you?

Knowing your records and knowing how to be prepared. All around, being prepared.

What's the worst experience you ever had at a gig?

[Chuckles]

Many! There are a lot of things! Girls taking off their clothes. People coming up to the turntable while I'm spinning and asking me questions about the groove. The power getting pulled. You know, that's typical.

Every DJ has probably had the power pulled in their set.

Fig. 6.2.4. Logic is equally comfortable mixing or playing in the band.

What kind of advice do you give to young DJs coming up in your neighborhood?

Younger DJs comin' up, if I have certain ideas, I just ask them, "Why don't you try this? See how this would sound."

But I also encourage them and give them props for what they're doing!

If they're trying to make something happen, but they're not hearing it in the place that would be cool to put it, I might want to correct them, just like you would correct a person holding a guitar or something. I may give some advice or pointers, if they ask.

And I try to encourage them to just be creative and go with their heart.

When you were growing up, did you have any DJs that gave you advice?

Oh, yeah, yeah. Africa Islam, Evil E. These were the guys touring with Ice T, when I was touring with Ice T in the alternative rock group. They gave me some pointers.

What kind of pointers did they give you?

You know, be yourself and just listen! Once you got that vibe happening, once you got those first few records going, just follow that flow and the rhythm all the way out. It was some positive pointers and "Keep it up, keep what you're doing."

I remember those words, speaking to those DJs. That made me feel good.

It wasn't like I was shying away from a scene. I was still representing a scene, which was Hip-hop, and I was part of the scene at the time. I was doing something with this alternative rock band, but it was handed down from these pioneer Hip-hop cats that I admired. That helped me continue onto what I'm doing now, playing with all these different musicians.

What do you think the Hip-hop pioneers like Flash and Bambaataa make of what you're doing, playing with bands?

I think they admire it. To see the whole thing that they started progressing to where it is now. People are just being creative and taking what they heard from them, and using it to the best of their abilities.

I mean, most of the stuff I'm doing is influence from those guys. I'm happy to be presenting it and turning other people onto it. They turned me onto it, and I just wanna give back and teach.

What do you make of the nu-metal scene and the turntablists in that scene?

I think that's cool! I look back to the day I started off with the alternative rock band. There was nobody really focused on what I was doing. Hip-hop had its thing, and rock had its thing, but nobody could see this new thing happening. When I played at CBGB, I would look at the people and they would hear what I was doing, and they couldn't relate, you know?

As time went on, the band and I kept getting more comfortable with each other, and people really noticed how I was playing my role and playing my part as a musician.

We recorded the album, and to this day, you could play the album over and over, and it sounds like a record you could make today.

That record was made in the 1980s, right?

Yeah, it was done in 1988 or 1989, somewhere in there.

Seeing DJs in bands today doing what they're doing, I'm happy! DJs are really not afraid to jam with musicians, and musicians are not afraid to jam with DJs. Everybody thinks it's cool, and they all can work together and be creative together. I think that's great.

Do you ever play with other DJs?

I've played with other turntablists. I've jammed with DJ Spooky, Kid Koala, DJ Greyboy, DJ Olive, and the Fifth Platoon, out of New York City. And everybody has different styles.

What's your favorite part of all of this?

Seeing people's reactions to the music. Traveling to all these different places and turning people onto something different. Playing with musicians in general. I get a buzz just playing with them. Everything jumps up another notch, and I think that's great.

My favorite part is going out and seeing other people's faces at the places I play, and just being able to do something that I love to do.

6.3

Faust and Shortee: Two Turntablists Mix Music and Life

Since meeting in 1995, Faust and Shortee have shared years of musical growth, turn-tables, a roof, and a life (Figure 6.3.1).

As a duo, Faust and Shortee have released numerous CDs, including the *Dream Theory* series and the *Hip Hop Mega Mix* series. Their tracks have aired worldwide on NBC, CNN, and ESPN with features on the *X-Games*, *Tony Hawk's Skate Park Tour*, *Hip-hop Nation*, *Headline News*, and the *World Cup 2002*. They've worked with the Red Hot Chili Peppers, DJ Irene, the X-ecutioners, Dieselboy, Crystal Method, KRS-One, Method Man, Dilated Peoples, Cut Chemist, the Skratch Piklz, Sasha, BT, and Paul Oakenfold. They even created a show featuring six turntables by teaming up with jungle phenomenon Danny the Wildchild.

In addition to their success as a duo, they've set many milestones independently: Faust's *Man or Myth* is considered by some to be the first solo turntablist album, and Shortee's *The Dreamer* was the first solo album by a female DJ/producer. In 1997, Shortee became the first and only female Fever/Buzz Battle of the DJs champion, win-ning the USA's largest techno, house, and drum 'n' bass DJ/mixing competition.

Faust was ranked number 5 of the top turntablists in the world by *Spin* magazine, and Shortee was recognized as one of *Urb's* "Next 100 Artists in 2001." She released mul-tiple instructional DVDs, and performed as *Playboy* magazine's one and only official club

Fig. 6.3.1. In 2001, Faust and Shortee became the first turntablist team to officially get married—to each other.

DJ to celebrate *Playboy's* 15th anniversary, at 52 exclusive high-profile events across the nation. The two also teach at Scratch Academy in Los Angeles.

When did you first brush up against DJ culture?

Faust: Herbie Hancock's "Rockit" and movies like *Breakin'*, *Wildstyle*, *Bodyrock*, and *Beatstreet*. The popularity of the break dancing culture was a new and exciting movement I could identify with because the music, dance, and art of the culture represented me in some way.

Shortee: When I was little, I used to make my own radio show tapes with some records and a little turntable, imitating the DJs I heard on the radio. At that time, I thought that was all a DJ was, until I saw Run DMC and Jam Master Jay on MTV. I asked for their first album for Christmas, but Santa misunderstood and brought me the Beastie Boys *License to Ill* instead. During high school, my friend Lark introduced me to the nightclub scene in DC, and that's when I first experienced the DJ as an act all on its own, rather than with a group. I played percussion and drums in traditional bands all through grade school and high school, and I didn't really look at the DJ as a musician.

That all changed when my friend Kiley took me to my first rave, and I started to get more into electronic music and DJs' mixed sets. In 1995, I saw Faust for the first time, scratching at a friend's house. That was when I truly began to become aware of the DJ as a musician and the DJ culture. I learned how to DJ myself, and I watched DJ battle tapes. The Skratch Piklz had just won the DMC, and they were helping to expose the turntablism movement around that time.

Who showed you your first moves?

Faust: When I was about 10 years old, my cousin in Puerto Rico showed me how to scratch. After that experience, I kept thinking about it all the time, until I got my first tables at age 13.

Shortee: Faust had seen me playing my drums at a few house parties with my punk rock band, Food. He encouraged me to learn how to DJ and scratch, because he thought I would pick it up really quickly. We were also dating at the time, and he had just moved in with his turntables.

I started playing house parties and gigs around the college area. We then moved to Atlanta in 1996, where I started playing more club gigs. Around that time, we met Shotgun, the DJ for the Goodie Mob, who introduced us to Craze, T-Rock, King James, and Klever. We all formed a turntablist crew called the Third World Citizens. I learned so much from practicing with those guys, especially Craze and Faust, while producing and touring together. But all of us learned from the X-ecutioners, the Skratch Piklz, and the Beat Junkies by being around them and watching.

Tell me about your professional DJ careers up until now.

Faust: I started playing Hip-hop and scratching when I was 13. I began my professional quest in 1996 with the aid from my partner Shortee, who gave me the most support. We started out doing all-night college house parties. Slowly, we built a name in this arena. We bummed rides to shows and played for cheap—sometimes for nothing.

Fig. 6.3.2. "Faust taught me the basics of mixing, scratching, and beat juggling in 1995"—DJ Shortee.

Fig. 6.3.3. "I saw the rave scene as a powerful stage for DJs, because it was the only place aside from competitions where the DJ was the show, the main act" –DJ Faust.

I went through many significant periods of growth as a DJ, but I guess the first real step up is when I got signed to Bomb Records to do *Man or Myth*, my first LP. The turntablist culture was just starting to pick up, and I came along at the perfect time to become a face in that movement. It was great promotion and opened many doors for press and booking opportunities. After that, I dropped *The Fathomless* EP with Shortee and Craze, followed by my second LP *Inward Journeys*. By this time, I could tell the popularity of the scratch-tablist scene was diminishing.

I also produced a lot of instrumental music, at the time.

We felt, to survive, we needed to establish ourselves as artists, not just DJs/turntablists, which was the stigma scratchers and jugglers got stuck with.

We signed a deal with Stray Records to release two projects of instrumental music that people could listen to and DJs could play. To us, it was a natural progression. We felt that we should be producing the stuff we would play and work toward live performances of original music. In a time when music is changing so fast, we wanted to be able to adapt to the time.

Shortee: I started DJing in 1995, and Faust was and still is my biggest influence, with Craze being my second. After I won the Fever/Buzz Battle of the DJs in 1997, I began to really look at DJing and production as a career. Faust featured me on his debut album. I did the *Fathomless* EP with him and Craze, and then I released my own solo album, making it the first and only turntablist album to be released by a female, to this date. On my album, I also produced all my own music, instead of going the remix route, like so many turntablists were doing at that time.

Faust and I started playing together regularly on four turntables, around that time. I've since released *Digital Soul* and *Satisfaction Guaranteed* as joint full length with Faust, along with tons of mix CDs, compilations, and singles. We've played all over the world performing sets of house, techno, Hip-hop, and drum 'n' bass.

Recently, we've begun performing with Final Scratch, and we have started various groups, performing with MC Supastition and Christopher Longoria. I've been producing various genres and various remix work.

What are your musical backgrounds, in terms of playing other instruments and music theory?

Faust: The turntable is the only instrument I have ever played! I never studied music theory, but I listened to a lot of jazz, salsa, merengue, rock, soul, funk, and Hip-hop, growing up.

Shortee: I started playing drums at age seven and played all through elementary, junior high, high school, and college, in concert, symphonic, jazz, and percussion ensembles, as well as marching bands. I took music theory classes in both high school and college. I also took piano, violin, and trumpet lessons when I was younger, but percussion was my main instrument. I play the drum set, snare drum, and many other percussion instruments, including marimba, xylophone, vibes, bells, chimes, timpani, and auxiliary percussion.

Do you read music?

Faust: I only have some general knowledge of music theory, and I can't read music.

Shortee: Yes, I read music. My sight-reading has gotten rusty since high school, but I can definitely still read well enough to understand and play a piece of music.

Did your musical background helped you learn turntable techniques more quickly?

Shortee: Definitely, without question. Understanding music gives you a huge advantage over the average person who wants to learn how to DJ. You already know how to read and play music. All you need to do is to learn a new instrument. Since you already understand how the music is laid out, you automatically know when and where to bring in your songs. All you really have left to learn is how to beat match.

If you can play an instrument, especially a percussion instrument, this task becomes very easy (Figure 6.3.4).

When learning a new scratch pattern, it's much easier for me to understand and break it down by thinking of it as written music. I learned certain basic techniques much faster than Faust because of this. Playing an instrument on stage wasn't a new experience for me, either, so I also never had any issues with stage fright.

Faust: Being exposed to music all my life and listening to Latin and jazz definitely expanded my knowledge of what was possible musically. Nothing is like hands-on experience, but being a good listener helped me learn the bulk of my scratching technique. Growing up, I didn't know anybody personally who could scratch, but I listened to a lot of records of DJs like Jazzy Jeff, Jazzy Jay, Cash Money, Code Money, Tat Money, Mr. Mixx, DJ Scratch, DJ Skill, Pete Rock, or just anybody who scratched. I used to listen to the scratch parts over and over until I figured out how they were doing their cuts.

Do you consider the turntable to be a musical instrument?

Faust: A DJ can use the turntable as a percussion instrument, and with certain turntables that have a higher degree of pitch control, even melodies can be created. That's why I consider it an instrument.

Fig. 6.3.4. "As a percussionist, I am used to using my hands to create rhythms and tones, so scratching to me is just using the turntable and mixer as another percussion medium"—DJ Shortee.

Shortee: The turntable is a percussion instrument because when you scratch or beat juggle, you are creating rhythms and tones with your hands, like you would if you were tapping a drum or playing an auxiliary percussion instrument, such as a guiro. The only real difference between these instruments is that the turntable has various sounds to choose from, like an electronic keyboard, whereas a drum is limited to the sounds of that particular drum.

The back-and-forth motion used when scratching a record is also very similar to the bowing technique on a violin.

The argument of whether or not the turntable is a real musical instrument is really a moot point, now. As DJ Radar has proven with his *Concerto for Turntable*, the turntable has been used to score a significant piece of music—not only as a percussive instrument but also a melodic one. Radar's *Concerto for Turntable* certifies and solidifies the turntable as an instrument in the history of music, because it's the first time the turntable's voice has been written out using traditional music notation. Any musician that knows how to read music can pick up the piece and play it on the turntable—provided that they're familiar with the technique of playing the instrument, of course.

How does your musical background affect your approach to the turntable and DJing?

Faust: Even in a mixed set, we try to incorporate the turntable as a control medium to layer and remix audio live. Our scratching and musical selection is definitely affected by our various musical influences. In a mixed set, we incorporate all types of audio from music to spoken word and sound effects. They all have different moods, and when combined with other pieces, sounds, or beats, we have more control over what energy we want to convey to the audience. Some stuff is multidimensional, in the respect that certain songs may get response from the crowd because of who the artist is, or they may recall some nostalgia when they hear a classic.

What's more difficult for you, scratching or beat juggling?

Shortee: They are two different animals. They both have easy and difficult aspects, respectively. It just depends on which one I'm practicing more at the time. Beat juggling has the added hurdle of finding the right beats to juggle together. Finding a sound to scratch is much easier.

Faust: Each has a specific set of rudimentary techniques. I don't think one can be more difficult than the other, because it all depends on how much time you sink into it. They're two different things.

How would you describe your current role in DJ culture?

Faust: We are pretty open-minded about music and have always experimented with everything, like Hip-hop, soul, funk, rock, house, drum 'n' bass, turntablism, etc. By playing multiple genres of music, and really trying to establish our names in each one of them, we have tried to bridge some gaps between the various markets and audiences. We continue to do that now by keeping our identity alive as DJs and producer/remixers in the scratch, Hip-hop, house, and drum 'n' bass scenes. We are DJs, but at the same time, we use the turntables to accentuate our unique voices and to create original music.

Fig. 6.3.5. "I personally spend more time scratching, so I know way more scratch techniques than juggle techniques" —DJ Faust.

How would you describe your live show?

Shortee: Our shows change all the time, depending on the event or where we are in our career. We definitely try to offer something different and stay away from playing anything commercial. We normally play on four turntables and have been doing so for five or six years. We are pretty in tune with each other on stage.

We rock mostly instrumental music and scratch through effects. We play many dance parties, so no matter what we play, our goal is to provide energy.

We also do showcases in which we do scratch routines and juggle.

How do you go about putting together scratch routines for four turntables?

Faust: We put together sets like one person normally would. Each of us has certain signature mixes or records we enjoy playing and think would fit in the set. Then we split the set by threes or fours. I'll play three or four, then Shortee will play three or four. We mix the music back and forth throughout the set. I'll mix to her, and she'll mix to me. While I am mixing, Shortee is layering or adding sounds, words, scratches, etc.— syncopated patterns that are mostly improvised and meant to add another level of progression to the music and vice versa.

What are your goals for "Faust and Shortee, the Live Show?"

Shortee: Live manipulation of audio and video. Our goal is to perform our original tracks and live remixes complete with video, which we can also manipulate via turntables.

Fig. 6.3.6. Shortee performing solo at Casa de Playboy.

What gear do you have in your studio?

Faust: We have two desktop PCs, one PC laptop, one Mac laptop, Soundforge 6, Acid 4, Logic, Fruityloops Pro, Reason. We love to play with software. We use a Korg Oasys PCI soundcard and MIDI Mapper, a Korg MS 2000 analog modeling synth, a Yamaha SP1200 drum machine, three Final Scratch systems, six turntables, lots of Vestax, Rane, and Stanton mixers, a Pioneer DJFX 500, Stanton DFX 1, and a Stanton DJ F1.

What's your favorite piece of gear, these days?

Faust: Final Scratch is probably our favorite piece of gear, right now. We have put together tons of remixes over the years that we intended on releasing, eventually, and now with the help of Final Scratch, we can play and manipulate them.

Final Scratch enables us to play things we probably would never get to play, such as tracks that we've remixed ourselves or original music that we can't afford to press up on dub plates.

This breaks down the barriers of DJing, and it opens so many doors. We can now concentrate on sounding the way we really want and work with pieces we want, as opposed to being limited to what we can find. It is also an incredible production tool, because it turns your turntables and mixer into a MIDI controller, and we can do things we've never been able to do before within our production. A basic example would be sampling our voices or a clip from a movie and scratching it in, on the fly.

What are the advantages and limitations to the system?

> **Faust:** Most DJs travel with a record box or a bag with sometimes up to 50 to a 100 records.
>
> With Final Scratch, your record bag is as big as the amount of hard-drive space your computer has!
>
> Another advantage is that it's compatible with MP3 files, and you can therefore bring thousands of tracks with you to a show. Pressing up vinyl is expensive, but vital to DJs who want to perform live with records. Final Scratch comes with time-encoded vinyl records, which look and feel like actual wax. The audio in Final Scratch can be played, scratched, needle-dropped, pitched, and basically manipulated just like a record. You can also use it as a production tool and make original songs and remixes, which you can test against store-bought records or test tracks on audiences and in live PA settings. It can also function as a controller.
>
> The downside of it is that you have to bring a laptop or desktop computer with you at all times. If the turntables don't play out of both channels or the contacts aren't good, it won't work. As a DJ on the road, you never know what kind of equipment you will have to deal with at a show, so you need to be prepared for the worst.

What are your goals for "Faust and Shortee, the Production Team and Recording Artists?"

> **Faust:** Some of our main objectives right now are remixing, licensing music, scoring for movies, videos, and other media. We are always working on all types of music and trying to push barriers in our productions, as well as shows. We plan to release some upcoming singles and EPs in the house and drum 'n' bass genres, as well as working on our newest solo projects and joint albums.

Are you interested in producing other artists as well?

> **Shortee:** Yes, we have worked with MC Supastition and are working on another project with him, currently. We also have a group with poetry slam champ Christopher Longoria called the Naturals, in which we paint audible pictures to his poetry via turntables.

Who do you consider to be the most influential people on the DJ scene today and why?

> **Faust:** QBert, Kid Koala, and Craze, because they are some of the most innovative DJs out there, who continue to consistently set new heights and standards for the DJ culture.

What does the future look like for Faust and Shortee, 10 years from now?

> **Shortee:** I just want us to be happy, healthy, and financially stable. Hopefully, we will still be performing, producing, and doing what we love to do together, 10 years from now.

6.4

On the Classical World's Radar

DJ Radar glanced up at the audience and wiped a bit of sweat from his brow before launching into a fury of crabs, waves, and uzis. The eclectic crowd at Carnegie Hall mingled dreadlocks, tattoos, and facial piercing with jackets and ties.

Behind Radar, 68 talented young classical musicians culled from Julliard, Manhattan School of Music, New England Conservatory, Eastman, Curtis, and Berklee, comprised the Red Bull Artsehcro named for their sponsor and the word "orchestra" spelled backwards.

The event was the world premiere of the completed three movement Concerto for Turntable, composed by Raúl Yáñez in close collaboration with DJ Radar (Figure 6.4.1).

I first met DJ Radar in San Francisco at Scratchcon 2000, QBert and Yoga Frog's historic event promoting scratch-music literacy. Radar had the first movement of his turntable concerto, I had my manuscript for *Turntable Technique: The Art of the DJ*; both works contained turntable music notation, so we had plenty to talk about. Producer Kurt Langer and director Doug Pray were running around shooting footage for what would become the movie *Scratch*!

Born in the Paradise Valley area of Phoenix, Arizona in 1977, Jason Belmont (AKA DJ Radar) became driven with taking turntablism where it has never gone before: to the symphony. But this classically trained percussionist has spent enough time in the DJ trenches to earn the respect of old-school spinners.

A member of the Bombshelter Crew since he was 13, Radar learned at the feet of Z-Trip and Emile what it takes to rock the house.

Fig. 6.4.1. DJ Radar's turntable concerto performance at Carnegie Hall.

Radar has also pioneered the technique of "live looping," which allows him to create turntable compositions in real time, while the audience looks on. His talents have been on display on records, as well as tours of Japan, Europe, and the USA, including tour stints for MTV and the DMC.

But on October 2, 2005, a few miles from the Bronx streets where Grandmaster Flash re-imagined the turntable as a musical instrument 25 years earlier, everything in Radar's background became a precursor for the event at hand. Wearing a traditional black tux, and without his trademark hat, Radar launched the turntable across a major threshold: as the soloist in an orchestral concerto, holding forth in the pre-eminent temple of classical music.

Radar's smooth introductory uzis gave way to Movement One's melodic themes shared between the string section and the turntable. Radar manipulated the pitch slider of a Numark TTX, using custom-pressed vinyl containing long string tones. Movement Two contained more precisely intonated melody playing by Radar using vocal samples. The third movement began with Radar performing live percussion looping, and developed into a showcase of his prodigious scratching technique, including impeccably even crabs over waves. His cadenzas started with controlled beat scratching and built into complex syncopated rhythms, climaxing in smooth uzi fades. A standing ovation punctuated the evening's accomplishment.

We conducted this interview on multiple occasions in New York:

What's the hardest part of being a turntable soloist with an orchestra at Carnegie Hall?

The hardest thing is the acoustics, especially of Carnegie; they're so sensitive; getting the velocities and the volume right. Also, no orchestra has perfect time. It's a constant challenge, but I love it, that's the fun part. The orchestra is like a big boat in the sea, so you've got to adapt quickly.

Looping with an orchestra is a challenge; it's really difficult to do. It's really dependent on the hall, and it changes every time. It's amazing how many times we've performed this and the things I've had to change. Like volume levels; in one hall, I'm always set right here, but when you go somewhere else, it's totally different. Different monitors work in different positions, with amplified stuff it's always different, I'm glad I had people there.

Did you feel a lot of pressure being the first turntablist to perform as the focal point of an orchestra in this particular space?

I was really shocked that right before I went on I was really calm, and usually I'm not like that. I'm usually kind of antsy with anxiety, but I was collected. I was just waiting; there's been so much going on this week but I had a moment of clarity there. It's a powerful experience.

My favorite part was hanging with the orchestra before we went on. Someone from the orchestra came up to me and asked, "Can we bob our heads?" And I was like, "Yes! Yes!"

When I first walked into the rehearsal space on the first day, the energy of the students was just so incredible, it gave me goose bumps. I took out the whole violin section for drinks the other night and we got to hang out, and it was awesome, they were so cool. They want me to play on their solo projects, and we already have more than a full plate of stuff. Someone pointed out "it's just like a whole lot of little viruses spreading out there."

Fig. 6.4.2. Rehearsing with the orchestra in New York.

How did you get into music and being a DJ?

I first got turned onto music through my music teacher at elementary school. I also got into music through my cousins, who all played instruments. My oldest cousin played drums and saxophone, and he was into Hip-hop. I was probably eight or nine years old when I first got into playing drums through him.

He also introduced me to Run DMC and Jam Master Jay, who became my biggest mentors. That's when I started hearing scratching, and I was just blown away. I got into Eric B and Rakim and all the classics, like UTFO and James Brown's Clyde Stubberfield. I just started collecting all these records, and that's what led me to DJing.

I got into it mostly for scratching, because it just really appealed to me. I always felt it was so rhythmic.

Did you play in high school band and orchestra?

Yeah, I played in all the junior symphony summer camps. I played a little bit of piano, but mostly percussion, like xylophone and bells. I come from a very heavy percussion background.

My teacher sometimes just wouldn't show up, so I would help prep the class, tune them, and conduct the pieces.

There was a really good snare player, when I was a freshman, who just drilled me with all the rudiments—same with my private teacher. He was a snare drummer for the circuses, and he drilled me on all the rudiments and flams.

During those same years, I started getting into clubs, because my sister, who is seven years older, knew everyone in the scene. I was young, and they would always have to sneak me in the back. I was always carrying records to get in.

What kind of clubs would you go to?

Mostly to underground raves. This was probably around 1990, and it was just starting in the Phoenix area. Phoenix was really small and really underground.

What DJs influenced you back then?

My sister introduced me to Eddie Amador, who is a huge house DJ now. He lived right by my high school. Pete Salis, who also lived there, was a Hip-hop DJ, and he took me under his wing and showed me some things. I would go over to their house, and he would show me how to scratch. He showed me the triple tear for the first time. I had seen it on videos, before, but I never actually saw someone do it live. I didn't even have turntables yet, so I would always go over there and practice.

Finally, when I was 13 or 14, I saved up money and got a Techniques turntable. I bought it off of a wedding DJ who had never even touched the pitch at all. This thing was pristine. I didn't have a mixer for a long time and used the volume on my stereo. That's how I'd scratch, but it wasn't ideal.

I did not have any money, in order to buy a mixer, so I went to Radio Shack and built a little switch for my phono. That's how I got started with the switch instead of the fader. I used to stay up late and listen to the mix shows on the radio, and go to raves. I was so obsessed with scratching. I would just study anything I could get my hands on.

Fig. 6.4.3. DJ Radar in New York's Times Square.

My sister also hooked me up with Emile, who probably influenced me the most of anyone in Arizona.

Who are some of your favorite turntablists now?

D-Styles is one of my favorites.

How did you get into the Bombshelter Crew?

One day, I was at a house party at this little underground club called the River Bottom Lounge, in Phoenix. This guy just walks in and just starts scratching, taking over the place like he owned it. I was like, "Who is this guy?" That's where I met Emile.

At the time, he was partners and best friends with Z-Trip, who was a big Hip-hop DJ, back then. I was really into his stuff and so Emile introduced me to him. They heard some of my scratching, and a couple of weeks later, they asked me to join their crew. They basically helped me get started (Figure 6.4.4).

Why should a DJ be in a crew?

Take Bombshelter, for example. Emile was doing all the raves and Hip-hop concerts. The collaboration between him and Z-Trip was interesting, because back then, you never saw those two worlds come together. This was a big jump for Z-Trip, because the Hip-hop crowd is more conservative. Emile comes from the rave scene, which is so much more liberal. It's all about the music and not about what kind of clothes you are wearing.

Emile believes that electronic music is only separated by tempos. He thinks that all of these genres like drum "n" bass, house, and techno are all the same. He used to show me how you take one record, and by speeding it up or down, you would have Hip-hop or break beats, and then techno or drum "n" bass. Z-Trip started breaking down some of his walls, and I started breaking down all of my walls. Emile just changed everything that I thought about electronic music.

Fig. 6.4.4. A flier advertising a gig by the Bombshelter Crew: Z-Trip, Emile, and Radar.

Emile was a techno DJ, but he would play anything on the board, and that really impressed me. Most other people would play house or Hip-hop all night, and they wouldn't even budge. Emile's probably the best mixer I've ever heard in my life. Watching him is what really got me into mixing. There's so much to mixing, and he's just the master.

We used to have battles. We would go through our collection, and he would hand me some records and say, "Okay, mix this!" He would always beat me.

[It blew my mind, how somehow, *Emile* would always find a way to mix *any* two records.]

Let's talk more in-depth about mixing. What are the most important issues to to keep in mind when you're learning to mix?

Knowing your levels, and learning how to separate your ears is fundamental. But the most crucial thing is knowing your records. Find out the BPM of all of your records, so you know the tempos when you're starting out.

The next step is balancing your headphone and your monitor levels. You don't want your headphone mix to be too loud to the point where you can't hear your monitor mix, and vice versa.

Do you usually have one ear in the headphones when you're mixing?

Yes. When I mix, I always start on the downbeat. If you know where the downbeat is, then you start listening for the snare, and matching the snare up to the other one.

If you know the snares of both records, it helps you distinguish between them. In many environments, and especially in your headphones, it's easier to hear the snare than the kick.

A lot of times, the bass frequencies are so low that they all blend together. That's why I think most people mix off the snare.

The hardest part is being able to distinguish the two records when you are slowly blending them in. I call that "separating your ears."

What do you use to BPM your records with?

I just use a stopwatch. My teacher in grade school drilled us with identifying tempos. She had me memorize tempos because I was the only snare drummer. Actually, my drum teacher told me to sleep with a metronome. So, I slept with a metronome for years!

What's a good sleeping tempo?

I would say, like 80 or 85. It actually helped me go to sleep! I must say that I was just blessed with really good music teachers throughout my school years. I was really into my teachers, and I got straight A's.

Tell me about your studies at Arizona State University.

I was very much interested in sciences already in high school, along with music. Music to me is so mathematical, and mathematics always made sense to me. That's why I started out at ASU as an astrophysics major. I was, and still am, really into studies of stars and gravity, protons, dark matter, and black holes. I could always make a mathematical connection between science and music.

Fig. 6.4.5. DJ Radar blending in with the locals on the New York subway.

I think the more humankind evolves, the more we will get to the point of recognizing everything around us as one big complete circle.

We're all made of atoms, and I think everything is connected together in the universe. I wrote a 40-page thesis on dark matter, in high school. The interest in this subject is the reason why I called my song "Antimatter."

Let's talk about practicing. Most people say to practice slowly and speed will come over time. Do you agree?

Actually, I always learned doing it faster. I found that scratching at fast tempos accelerated my learning curve. Instead of practicing it slow and speeding up, I was practicing fast and trying to get it from sloppy to not sloppy. At the end of my practice session, I would go down to a slower tempo, and I had more control so that I could figure things out easier. When you scratch faster, I think it opens up your muscles a little bit. It's all about memory recall. Your brain is much faster than your reflexes.

How do you mark your records?

I don't mark my records at 12 o'clock, but rather, toward the needle. It helps me, when I scratch, because I'm always looking at my needle as a reference.

Describe to me your style of playing.

My DJing is more like Bombshelter, based on the philosophy of collecting all kinds of music. Anything that I feel is going to make the crowd dance. I'm really into electro, these days. But whatever genre it is, if I really like a song, I'll try to find a way to work it into my set.

It depends on where you are DJing on a particular night. There's so much to rocking a crowd from just a mixing standpoint. There's so much to learn, and it's taken me 10 years to really get a firm grasp on it.

You're always learning new crowds. Whatever you expect can change so suddenly, so you should always be prepared and bring all types of music. I do so many different types of events; you just never really know how to prepare for it.

I like to start the night with real slow tempos, in order to create an intro type of vibe. Throughout the night, you speed up the tempos and add more energy, and then certain times you can drop it down. The simplest way, and this is what we do with Bombshelter, is to start slow and build the tempos and energy. We keep going and going, faster and faster, throughout the night, over a three-hour period.

But you can also go back and forth between building it up and dropping it down.

The problem is that you can't sit in your bedroom and plan out a set. You have to work with the crowd, pick up on their moods and improvize as you go.

You have to balance out between playing your set and improvizing.

If you are in a situation, however, where you are going against a DJ that is really fast and has a lot of energy, you can kick it off by picking up the high energy and then do the opposite.

Another important aspect is transitions. With mixing, you have to know your intros and outros. I try to keep my records in groups of similar tonal vibe.

Fig. 6.4.6. "I could always make a mathematical connection between science and music"—DJ Radar.

Let's talk about Concerto for Turntable. When did you first become interested in bringing the turntable into an orchestral setting?

This is just an idea I always had. But before making it reality, I had to educate myself and do a lot of research. I needed to develop a notation system—a way to communicate with composers and conductors so that they could follow what I was doing.

When I started working on the song "Antimatter," I wanted to try a new approach where I would first write out the song and then perform it. I started notating structures of things I wanted to do with rhythms, changes, and motives. There was one part in "Antimatter" that was just so complex that I didn't know how I was going to do write it out.

A friend introduced me to a jazz piano player at ASU named Raúl Yáñez. We developed this amazing collaboration, where he would teach me about notation and give me many ideas on how to notate the most difficult passages.

We would study classical scores together and try to adapt some of the notation to the turntable. I started creating these different articulations for all types of scratches.

When I told him that my ultimate goal was to write a piece for an orchestra and a turntable, he got really excited. There were already DJs that had played with orchestras, but I wanted to do something new: all scratching.

We hit a point where we got a little bit frustrated in the process and decided to take a break. Later, during a trip to Japan I saw this prototype turntable: the PDX 2000 by Vestax. This turntable had an ultra pitch, which gave it over an octave to work with! I brought one back with me. Our original idea was just to score "Antimatter," but once Raúl figured out what that turntable could do, we decided to write the concerto—what later become known as the Concerto for Turntable.

Had he ever written a concerto before?

No. What's crazy is that he is a jazz composer. The next hard step was to convince an orchestra to perform the concerto. We didn't know anyone in the classical domain, so we just started e-mailing various orchestra directors. At first we didn't get any responses—which did not surprise us.

Finally, I got an e-mail back from Joel Brown, who is a conductor at ASU. He was a masters conductor student, at the time, as well as the maestro in charge of all the ASU Symphony Orchestra. He expressed interest in having the piece performed at the annual winter concert series. I was totally excited! In preparation for that, Raúl and I spent nine months just writing the first movement. It was a lot of trial and error.

The first movement is about 12 minutes, but it's pretty intense and intricate, with a lot of melodic and rhythmic passages.

Does Raúl write on the computer?

Yes, we did everything on Finale. We used some synths, like the Roland and the Virtuoso, in order to roughly hear the composition. But we really learned a lot in the rehearsals. We had about four rehearsals with the full orchestra, and we had a couple of sectionals. We realized that some things did not work out, and we had to go back make changes in the percussion, for instance. We had a lot of people help us out, making corrections and improvements, because there is so much to concertos and orchestra pieces. After all, it was Raúl's first orchestral work he had ever written!

The first rehearsal was awesome. All the directors were there. I was nervous, but it was the best experience. I couldn't believe I was scratching with an orchestra! We had a video crew come in, and they did a documentary on it. They followed us all around.

It was great to bring these two worlds together, the club subculture and the classical world.

They don't know about each other, and now they are being united as one. I'm just trying to educate people and trying to bring people together. That's what music should be about.

How did you amplify the turntable so that the orchestra could hear you?

We had nine speakers. Some, in the back, for the percussion, in the basses, and then we had some in the middle. We had two upfront, and I had a monitor. Only the conductor did not have a monitor because it was loud enough (Figure 6.4.7).

Tell me about the night you premiered the first movement.

It was the most amazing experience.

We didn't even promote this event and 3500 people showed up. It was all just word of mouth. The night of the show, I was sitting backstage, thinking to myself that this was one of my dreams come true. It was the most diverse crowd I've ever played for in my whole life.

People of all ages and ethnic backgrounds were there. That alone was the best experience for me.

I was really nervous, but I walked on stage, and my mind was totally cleared, just seeing all the people and feeling the energy. I never experienced silence as being so loud! I could feel the anticipation. And the roar from the crowd when I did the cadenza was just overwhelming. The whole place just erupted. People clapped in the middle of this orchestral piece. It was really funny, because most people don't know the etiquette. That was awesome.

At the end of the piece, people were yelling "Encore!" So, we ended up playing the piece over again from the beginning.

People went just as crazy! It was nuts. It was wild. It totally turned into this rock concert. The whole staff of the department was there, and they were just so blown away by how many people turned out. They completely had a new respect for this new culture.

Are you going to keep working with Raúl?

Oh, definitely. We just started this little jazz duo. A lot of our ideas came from improvizing. He's from a total jazz background and an amazing piano player.

Now, we can experiment and get a little deeper into some abstract rhythms and a whole different genre. Raúl is incredibly gifted. He's got some amazing ideas.

He had a jazz trio that played a piece that was written on the turntable. I think that turntablism is related to jazz in a lot of ways.

Fig. 6.4.7. DJ Radar and the Arizona State University Symphony Orchestra at the premiere of the first movement of the turntable concerto.

What do you think about the viability of the turntable as a melodic instrument? Do you think more people are going to see it that way?

I think we're just getting into the realm of that. There's definitely a lot more to do with the melodic capability. I mean, obviously, the rhythmic stuff has expanded so much, but a lot of turntablists haven't realized what they are doing melodically. I'm really trying to push it to make it a melodic instrument as well.

What areas do you personally want to explore in the future?

We learned so much doing the concerto, now it's back to the drawing board. We've already started piecing out the next concerto. I don't know if we're going to call it "concerto," but it's definitely going to be the next turntable and orchestra piece; it's going to be like the best of the best. Over the last four years we've learned incredible amounts of information that we need to get back to and revisit. We've just been putting stuff in the safe. The next one is going to be 10 to 20 times better, because there's so much we've learned; how to not confine ourselves.

I'm going to be doing this orchestra work for a while. Jazz is another genre we're working with. I also want to learn more instruments. I'm picking up the violin, right now. I think it's just going to help my composition more.

I'm going to keep working on the academic aspect of turntablism, because there are still a lot of walls to break down. I want to be educating people and sharing.

I want to have a huge arsenal of projects I have worked on. Ultimately, maybe teaching or working on film scores. I know, 30 years from now, I'll be doing music. I'll be scratching forever. It's changed my life, you know? It's just not going to go away. I love it too much.

Someday, I want to DJ on Mars or the moon. I'd love to play in outer space. Get me to the space station, and throw a party!

PART II

The Tools

DJs are always pushing at the limits of technology.

David Mancuso's high standards of fidelity led to new designs by audio designers Richard Long, Alex Rosner and Louis Bozak, setting the stage for decades of innovation in the field of DJ Tools.

The Adventures of Grandmaster Flash on the Wheels of Steel was created entirely with a basic two turntable and a mixer set-up, pushing the limits of this technology past anything the manufacturers could have imagined. Aspiring DJs with more time than cash have always pushed their equipment far past its intended use; getting dangerously good at looping and sampling with nothing more than a cassette deck with a pause button.

Drum machines, electronic effects, and multi-track recorders were embraced by DJs early on; *Planet Rock* is one prime example of the results.

In recent years manufacturers have been giving the craft of remixing (both live and in the studio) a lot of attention. The following chapters will explore the evolution of these tools and the current state of the art, and is meant as a guide to help you determine what's best for you.

7

Turntable Tools

In recent years, the turntable has undergone a revolutionary transition from utilitarian tool to musical instrument.

While the earliest versions of stylus-in-groove machines did not lend themselves to the aggressive manipulations that current machines can handle, they did have one distinct advantage: wax cylinder machines could record as well as playback!

The recordings were noisy and playback volume extremely limited by today's standards, but all Edison wax cylinder machines came with both a recording stylus and a playback stylus for converting sound waves into mechanical energy and back again (Figure 7.1).

Phono Cartridges

The phono cartridge is the most critical and sensitive component on the turntable. Like a microphone, the cartridge is a transducer, which is a device that converts one kind of energy into another. While the microphone converts physical energy, the compression and rarefaction of air molecules known as sound waves, into electrical signals, the phono cartridge converts a solid representation of those sound waves, the grooves on a vinyl disc, into electrical signals.

On the original wax cylinder machines, both the record and playback stylus used a diaphragm to vibrate, in much the same way that a microphone does (Figure 7.2).

The horn focuses the sound waves onto the diaphragm on the way in, vibrating the stylus, which cuts grooves in the rotating cylinder. When replaced with the playback stylus, which tracks the grooves rather than cutting them, the

Fig. 7.1. A musician recording onto a wax cylinder machine.

Fig. 7.2. With the horn removed, the diaphragm of the wax cylinder machine becomes visible.

Fig. 7.3. Shawn Borri of the North American Phonograph Company listens intently to the vibrations of the stylus on his Edison wax cylinder machine.

diaphragm vibrates in sympathy with the grooves already on the cylinder. The horn amplifies the vibrations of the diaphragm acoustically.

Flat disc records are mastered on cutting lathes that cut grooves onto a master disc, which is then electroplated with a metal alloy. These are used to create negative "stampers" to be used in hydraulic presses to mold the LP discs at the manufacturing plant.

The modern phono stylus converts mechanical energy into electrical signals, sending them first to a pre-amplifier, then to amplification, and finally to speakers, where they are converted back into mechanical energy in the form of vibrations, which create waves of air pressure that we perceive as sound.

The modern stylus replaces the diaphragm of the wax cylinder machine with a cantilever, a very thin, stiff piece of metal, as the vibrating device. Like a microphone, the modern cartridge also uses a magnet and coil to convert mechanical energy into electrical signals by creating fluctuations in a magnetic field. The moving magnet cartridge is the most widely used style of cartridge, it also boasts a higher output signal than the other option, the moving coil cartridge.

Here's the way it works:

1. As the stylus traces the grooves in the record, it vibrates the cantilever, creating mechanical energy.

2. The cantilever is connected to the magnet, which moves between the pole pieces.

3. The movements of the magnet create a flux current, which travel through the pole pieces in the form of magnetic energy.

4. Coils wrapped around the pole pieces convert this magnetic energy into electrical signals, which are sent through the head shell leads (Figure 7.4).

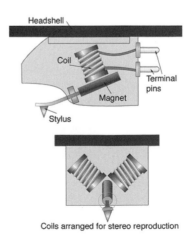

Fig. 7.4. The moving magnet cartridge, with its components labeled.

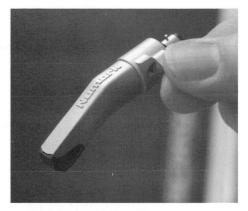

Fig. 7.5 and Fig. 7.6. The Shure M44 (left) is a standard mount cartridge, while the Tony Tone signature model (right) is a p-mount cartridge design.

To reproduce stereo records, two coils are used, positioned on either side of the magnet: one for the left channel and one for the right. The stylus reading a stereo record will respond to both sides of the groove (left and right), vibrating horizontally as well as vertically.

Cartridge Mounts

There are two radically different designs for cartridges, the standard mount and the p-mount (Figures 7.5 and 7.6).

Standard mount cartridges connect to a separate head shell with small screws and

Fig. 7.7. An inexpensive belt-driven mechanism.

nuts, and make connections via color-coded leads. Some scratch DJs angle the cartridge to more squarely place the stylus in the groove when using an S tone arm, emulating the angle of a straight tone arm. P-mount cartridges come with their own head shell, minimizing set-up time and effort.

Driving the Spinning Wheels

Modern turntables can be divided into two categories: belt drive and direct drive.

Belt drive turntables transfer torque from the motor to the platter by way of one or more belts, like fan belts under the hood of a car. Most inexpensive turntables that come in "DJ in a Box" packages are belt drive (Figure 7.7).

Using slip mats to effectively decouple the record from the platter is especially important when using inexpensive belt drive turntables for scratching and beat juggling. While most DJs start out on inexpensive belt drive decks, if you're serious you'll want to move up to direct drive decks made for more professional DJ applications when you can afford to do so.

Interestingly, there are also extremely sophisticated, audio-file belt drive turntables, some of which use as many as three motors and cost upwards of $10,000 (Figure 7.8). A great deal of innovation has gone into the design of these high-end machines in an attempt to recreate the exact motion of the cutting lathe that made the master disc. These precision instruments prove that belt drive systems are certainly not inherently inferior to direct dive decks.

The Old School 1200

Technics, a successful electronics company located in Japan, introduced the direct drive system in 1969 with the SP 10 turntable. With the direct drive (or DD) system, the motor is

Fig. 7.8. The Clearaudio Maximum Solution Turntable is handmade in Germany. Current price for this belt drive "Wheel of Acrylic" is around $6500, which does not include the tone arm.

coupled directly to the platter, turning at the same rate as the record. Using a low-speed, high-torque motor with no additional parts to transfer torque from motor to platter, direct drive systems feature quick start-up, low wow and flutter, low rumble and excellent rotational stability.

The massive SL-1100 came out in 1971, and introduced many features that would wind up on the eventually ubiquitous SL-1200. Originally launched by Technics in 1972, the 1200 featured an "S" shaped tone arm and a rotary pot pitch control. The deck quickly became an industry standard at radio stations and nightclubs, and the deck to aspire to among mobile DJs.

In 1979 the ±8 percent sliding pitch control replaced the rotary pot on the SL-1200 Mk2. The SL-1210 Mk2 replaced the silver colored chrome body with black chrome (Figure 7.9).

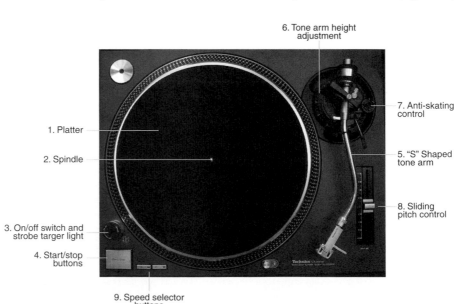

6. Tone arm height adjustment

7. Anti-skating control

1. Platter

2. Spindle

5. "S" Shaped tone arm

8. Sliding pitch control

3. On/off switch and strobe targer light

4. Start/stop buttons

9. Speed selector buttons

Fig. 7.9. Technics SL-1210 Mk2 direct drive turntable.

The Technics direct drive SL-1200 Mk 2 and Mk 3 series turntable was the workhorse instrument for DJs during the pivotal decades of the 1980s and 1990s.

Virtually unchanged since the late 1970s, the Technics "Wheel of Steel" had the torque to go from 0 to 33 1/3 in 0.7 seconds.

Let's take a look at the parts of the 1200:

1. **Platter:** The "wheel of steel" is actually made of die-cast aluminum, and spins by connecting with the motor underneath. The record is placed on top of either a slip mat (to slip cue, scratch, or beat juggle) or a rubber grip mat (for better coupling between the record and the platter).

2. **Spindle:** Keeps the record in place by fitting through the hole.

3. **On/Off switch and strobe target light:** The switch powers up the electronics, while the red light illuminates the dots on the side of the platter. When the speed is locked in at exactly 33 1/3 or 45 rpm, the dots will appear motionless. Slowing down or speeding up the platter will cause the dots to appear to be moving clockwise or counterclockwise, respectively.

4. **Start/Stop button:** Starts and stops the platter spinning.

5. **"S" shaped tone arm:** Positions the stylus slightly in front of the curve in the record groove. S-shaped tone arms have a natural inward pull, which can be off-set with anti-skate.

6. **Tone arm height adjustment:** Raises and lowers the tone arm base assembly, affecting the incline of the tone arm and the gradient of the stylus in the record's groove.

7. **Anti-skating control:** Adjusts the tension of a spring to counteract the natural tendency of the tone arm to be drawn toward the spindle: too much of it and the stylus will have a tendency to jump back, too little and the stylus will jump forward.

8. **Sliding pitch control:** Lets you slow down or speed up the record by plus or minus eight percent ($\pm$8%), which works out to a little more than a half step up, and a whole step down.

9. **Speed Selector buttons:** Chooses between the base speeds of 33 1/3 and 45 rpm (revolutions per minute).

While the 1200s included a basic feature set compared to the competition, their heft and torque made them the top choice among professional DJs for over two decades.

Techniques has updated the SL-1200 line with the SL-1200Mk5 and the SL-1200M5G, which increases the pitch control range to plus or minus 16 percent ($\pm$16%).

The New School TTX

In 2002, Numark introduced the TTX turntable, offering an expanded array of modern features, hefty steel and rubber body, and enough torque to reach 33 1/3 rpm in under 0.2 seconds (Figure 7.10).

Let's take a look at what's different and what's the same on the TTX:

1. **Platter:** Traditional.

2. **Spindle:** Traditional.

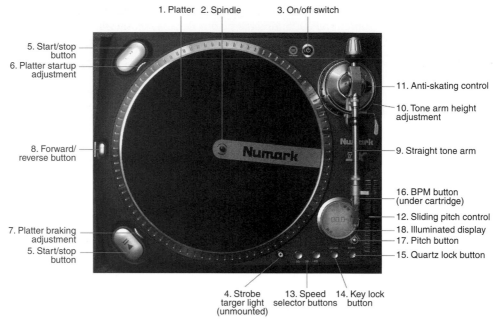

Fig. 7.10. The Numark TTX direct drive analog/digital turntable.

3. **On/Off switch:** Glows blue when on, separate from the target light.

4. **Strobe target light:** Removable (and aim-able) aluminum cylinder houses a bright white LED, illuminating the platter's markings in the traditional manner.

5. **Start/Stop buttons:** There are two, giving the turntablist more options in various playing modes and positions.

6. **Platter start-up adjustment:** Thumb rheostat which adjusts the start time of the platter from instant to several seconds, used for musical effect.

7. **Platter braking adjustment:** Thumb rheostat which adjusts the time it takes for the platter to stop, from instant to several seconds, also used for musical effect.

8. **Forward/Reverse button:** Instantly changes the direction of the platter, allowing you to play records backwards.

9. **Straight tone arm:** Places the needle tangent to the curve in the record groove, resulting in no natural inward or outward force on the tone arm. Reduces skipping, making it arguably better for scratching than the "S" tone arm. The TTX comes with both straight and "S" tone arms, which can be switched without tools.

10. **Tone arm height adjustment:** Traditional.

11. **Anti-skating control:** Traditional.

12. **Sliding pitch control:** Lets you slow down or speed up the record by plus or minus 50 percent ($\pm$50%), which works out to a perfect fifth up, and a full octave (twelve half steps) down.

13. **Speed Selector buttons:** Chooses between the base speeds of 33 1/3 and 45 rpm. Pressing both simultaneously will increase platter speed to 78 rpm.

14. **Key Lock button:** Locks the key in place digitally, allowing the user to adjust the tempo independent of pitch.

15. **Quartz Lock button:** Locks the motor to the internal vibrating quartz crystal, providing the most accurate rotation at 33 1/3, 45, and 78 rpm, and disabling the sliding pitch control.

16. **BPM button:** Shows the estimated BPM read-out in the center display. Holding the button for two seconds will recalculate the BPM estimate.

17. **Pitch button:** Toggles the pitch range between 10, 20, and 50 percent. Holding the button down for two seconds will set the pitch range at 8 percent to emulate the Technics 1200.

18. **Illuminated display:** Provides information on pitch range, RPM, BPM, key lock, quartz lock, start time, break time, and platter direction.

The TTX is actually a hybrid analog/digital turntable, with an S/PDIF digital output (Sony/Philips Digital Interface Format), which lets you connect digitally with a computer's digital audio interface, a CD recorder, digital mixer, or many other digital devices. Internal digital processing is what allows the user to change the tempo of the record without changing the pitch. This also makes it possible to change the pitch of the record, lock the new key in place, and return to the original tempo.

Another innovation is the analog outputs, which are switchable between phono and line level. Typically, phono outputs need to pass through a phono pre-amplifier which send the signal through the RIAA (Recording Industry Association of America) EQ curves, allowing the grooves in the record to be more manageable. The TTX has this circuitry built-in, making it possible to plug directly into any line-level input. Numark also decided to internally ground the TTX, negating the need for the ground wire which is a staple (and a hassle) on most turntables.

To accommodate both club and battle style mixing positions, the pitch fader and button cartridge are interchangeable, and the orientation of the display automatically changes when you switch the position of the pitch fader and the buttons (Figure 7.11–7.13).

Fig. 7.11, Fig. 7.12 and Fig. 7.13. Switching the position of the pitch fader and the buttons on the TTX.

New Turntable Designs

The Vestax QFO represents new thinking in turntable design. Developed in cooperation with DJ QBert and Thud Rumble, the QFO is a combination turntable and mixer, designed to appeal to the turntablist who approaches the turntable as a musical instrument (Figure 7.14).

Innovations include an anti-skipping tone arm that uses a spring to keep constant pressure on the record at any angle, and a 180-degree spin slide pitch control (±60 percent) located next to the platter, designed to make it easier to use the QFO to play melodies with tone records.

Companies continue to innovate with the lowly record player. In recent months, USB turntables, hybrid CD, and vinyl turntables, even turntables shaped like guitars have come to market. While the record industry wrote off vinyl records decades ago, the appeal of these devices remained.

Fig. 7.14. DJ QBert playing his invention, the QFO.

8

CD, DVD, and Hard Drive Decks

CDs offer DJs many advantages over vinyl. The reduced size and weight of both the playback medium and equipment made the CD, and now DVDs and Laptops, the preferred formats for most mobile DJs, who need to be ready to fill every request or risk facing the legendary wrath of the bride, or worse, the *mother* of the bride.

Many club DJs and turntablists stay with vinyl, citing vinyl records' temporal characteristics, distinctive sound, visual appeal, and the techniques one can execute using turntables that just can't be duplicated on CD players.

Hardware manufacturers have been hard at work to bridge these gaps.

DJ CD Players

The current generation of professional DJ-oriented CD players are feature-laden devices that make a strong case for mixing and scratching off of compact disc.

Dual-well CD players have been standard equipment for most mobile DJs, and for good reason. They conveniently feature two CD players in one rack-mountable unit, with a separate dual controller. These controllers range from being functional, to offering many impressive options, depending on the make and model. The jog/shuttle wheels on the CDN90 by Numark can be used for real-time scratching, as well as high-speed rotary track access. Dual pitch control faders can be set to ±6, 12, 25 or 100 percent, which allows you to go up an entire octave or grind down to a complete standstill (Figure 8.1).

In addition, there are multiple digital effects built-in, a key lock feature, adjustable start-up and braking speeds, S/PDIF coax digital outputs, a Musical Instrument Digital Interface (MIDI) input and output for synchronizing to MIDI compatible drum machines, keyboards and sequencers, and the ability to save up to 3000 cue points in memory.

CD players like the CDN90 also allow the user to easily loop sections of a CD live, a practice similar to extending breaks. Also made easy are techniques known as *stutter starts* or *stuttering*, somewhat akin to *cutting* on a turntable, though a bit more digital in sound and execution.

Fig. 8.1. The CDN90 dual-well CD player from Numark features built-in beat detection technology as well as auto-mixing and synchronization features.

The inexpensive and flexible nature of burning custom CD/Rs has helped mobile DJs cut down on the number of discs needed to play a gig. Rather than bringing along an entire album just to play one song, frequently played tracks can easily be combined on custom-burned compilation CDs. This assumes that the DJ CD player is CD/R compatible; most new decks are, however many older decks are not.

While the digital scratching function works well and sounds good, the jog/shuttle wheels on the CDN90's controller are obviously not made for turntablists looking to replicate the feel of vinyl, but for mobile DJs looking to incorporate a bit of scratching into their sets.

Music As Compressed Data Files

CDs ushered in the era of digital music playback in the 1980s, employing Pulse Code Modulation or PCM, which is simply digitally sampling a sound many thousands of times a second. What's become known as *CD quality* is audio sampled 44,100 times per second in 16 bit words (44.1 k/16 bit), a resolution capable of reproducing a little more than the audible frequency range of human beings; generally thought to be from 20 to 20,000 cycles per second.

In the decades since CDs were introduced, computer scientists have devised ways of making digital audio files much smaller than the 10 mega bytes (MB) per stereo minute that straight PCM requires. MP3 and MP4 employ data compression algorithms that pack pre-recorded digital sound into vastly smaller files, while seeking to retain as much audio quality as possible. A four-minute song takes up only about 4 MB of space encoded in MP3, as opposed to about 40 MB for a straight PCM file.

While many audiophiles and sound engineers revile these "lossy compression schemes" as being detrimental to sound quality, the formats have caught on with both the public and many DJs for the overwhelming increase in flexibility they offer.

In Chapter 9 we'll explore the practice of mixing off of a laptop computer using of compressed audio files, here we'll consider another option: mixing MP3 CDs. MP3 CDs are CDs that contain MP3 audio files rather than PCM audio files. Not all CD burners will burn MP3 CDs, and not all CD players will play them, but most burners and players manufactured since 2001 include this capability. The main advantage: you can fit approximately 11 hours

Fig. 8.2. The Numark CDN95 is a professional MP3 compatible dual CD player with real-time pitch adjustment of ±100 percent, key lock, an integrated beat counter, and built-in DSP effects.

of music (about 130 pop songs, or 10–12 albums) on a single CD/R using MP3-encoded files (Figure 8.2).

Another advantage that MP3 files have over CD audio is support for ID3 tags. An ID3 tag is a data container within an MP3 audio file that usually contains the Artist name, Song title, Year and Genre of the audio file, making it much easier to find a track on an MP3 CD.

CD Turntables

It was just a matter of time. Combining the art form of the turntablist with the advantages of the CD is one of those concepts that became inevitable. The falling price of digital memory buffers, significant advances in time expansion and compression algorithms, and the imagination of some really smart people all combined to set off a firestorm of change in the DJ world.

Pioneer got off a shot that was heard around the world with the CDJ-1000 in 2001. Established Hip-hop DJ Jazzy Jeff and Goth turntable virtuoso DJ Swamp were among the earliest to publicly embrace the new tool (Figure 8.3).

So, what is the difference between a DJ CD player and a digital turntable? The main difference is the interface. Digital turntables incorporate a touch-sensitive jog/shuttle wheel that emulates the platter of a vinyl turntable. Cueing, beat matching, baby scratches, drags, stabs, cuts, transformers, flares, uzis, pats, lasers, beat juggling, and live drumming techniques can be performed in approximately the same way one would perform these techniques using vinyl.

There is no needle to skip out of the groove; what you're scratching is digital audio from the CD being held in a memory buffer. This has many advantages, including the ability to scratch your own sounds (or beats, or vocal tracks) simply by burning them onto CD, rather than going through the trouble, time, and expense of having them pressed onto vinyl or acetate.

Fig. 8.3. The CDJ-1000 Digital Turntable.

The Pioneer CDJ-1000 features visual displays that provide information on most of the deck's functions. This includes a visual depiction of the relative location of the cue or sample that you are scratching, which is a digital LCD version of a sticker or a mark on a vinyl record. It also depicts a rather rudimentary waveform across the top of the player, which signals you when the track has 30 seconds and then 15 seconds left to play, by flashing on and off (Figure 8.4 and Figure 8.5).

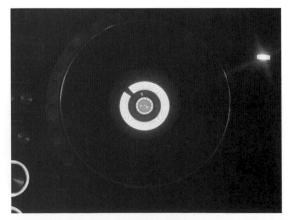

Figs. 8.4. & 8.5. The visual location interfaces of the CDJ-1000; the platter and the waveform display.

If you told me five years ago, "you'll be using CD turntables," I'd have said, "get out of here—I'm never using CD turntables!" But they were able to make a mechanism that was similar to the turntable, and that machine (CDJ-1000) has given me so much more freedom, it's actually made me more creative. For the next X-ecutioners album, that's all I'm using. I haven't used (vinyl) turntables yet.

—Rob Swift. Rob still uses vinyl when he plays live.

Another advantage the CDJ series brought was easy looping, and the ability to specify three "hot cues" (specific spots in the record), which can be instantly recalled to create new arrangements, or to facilitate "slam mixing" all on a single digital turntable.

Another entry into the CD turntable game is the Denon DN series, which added a 7-inch spinning platter on the top of the player, and what Denon calls "Alpha Track," which lets you play multiple tracks off of one CD at the same time. Technics also makes a digital turntable, the SL-DZ1200, which is supposed to emulate the SL-12000, but has a space age looking clear piece of plastic which is kept in place with some loose fitting rivits. I haven't met anyone yet who thinks it feels like its analog counterpart. The coolest thing about the SL-DZ is its ability to play four samples simultaneously.

The deck that did the best job of bridging the gap between the vinyl turntable and the CD player is the CDX by Numark (Figure 8.6).

The CDX feels so identical to scratching on a turntable, I've literally seen famous scratch DJs manipulate samples on the CDX for a couple of minutes before they noticed there was no tone arm, and no needle in the grooves. The deck features four scratch modes, and several onboard digital effects, including phasing, filters, auto-decimation, beat-synchronized echo, panning, chopping, and auto panning. The built-in Beatkeeper technology detects the tempo of the track playing, letting you choose echo and panning effects in eighth notes, quarter notes, or half notes, and will even synchronize automatically with a track playing on another CDX or a MIDI capable machine or sequencer.

All the CD turntables we've looked at here have adjustable start-up and braking speeds (similar to the TTX1), and the ability to play CDs backwards as well as forwards.

Fig. 8.6. The CDX shares the same footprint and high-torque motor with the TTX turntable, but manipulating the 12-inch vinyl on top controls sound from the CD inside.

Hard Disc Decks

While laptop DJ applications and CD turntables have been getting all of the attention lately, there are Hard Disc DJ devices that offer some of the best of both worlds.

Both Denon and Numark offer controllers that look and act like dual-well CD players, but play music off of internal hard drives. These machines can store tens of thousands of songs, and pretty much negate any reason for dragging CDs to a mobile DJ gig.

Numark has also adapted the technology developed for the CDX into the HDX, a hard drive equipped turntable/CD unit. The hard drive can be loaded by a computer via USB 2.0, and even has an onboard Gracenote™ CD database to automatically recognize and tag audio CDs, which can be ripped from the slot loading CD drive. Finally, the HDX has the ability to record as well as play back, making it possible for DJs to record their live performance on the deck they are performing from.

Vinyl Comparison

So, why does anyone still want to use vinyl? I'm hooked on both vinyl and digital, so let me try to explain why I still love to spin wax.

One of the few ways that vinyl still outperforms CDs is in the speed of changing records. You don't have to wait for the CD player to spit out your CD after you press the eject button, or wait to load in the new CD. If you're well-practiced, changing records can be both swift and elegant. Of course, if you have thousands of MP3s on a CD, you may not need to switch CDs.

It's also more visually interesting for the audience to see someone selecting, changing, and cueing up vinyl records than merely pressing buttons on a CD player.

In addition, DJing can be a performance art. When you take the element of risk out of a performance, it can't help but be less interesting. When there's no chance of the needle skipping out of the groove, you've essentially removed some of the danger or drama from the performance. True, it may save the performer and the audience from the pain of a botched presentation, but the same is also true of a vocalist lip-syncing or using an auto-tuning device in live performance. How much more exciting is the high-wire act when the artist is working without a net?

On the other hand, mobile DJs are often in a situation (weddings, anniversaries, sweet sixteen parties, bar mitzvahs, etc.) where the point is to lavish attention onto the participants, deflecting it away from the DJ. Further, while the list of turntable techniques not possible on CD turntables grows shorter, the list of new techniques that these decks make possible grows larger with every new generation of hardware.

Fig. 8.7. DJ Logic hits the crates, looking for the perfect beat.

Computerized DJ/ Remix Tools

When it comes to hi-tech tools for DJs, there are plenty of hardware and software applications out there for playing live, remixing, making beats, recording, sequencing, and even turning your analog turntables into laptop interfacing, digital audio and video machines.

In fact, there are so many choices out there it would take an entire book to cover them all, so we'll focus on some of the most popular software and hardware chosen by leading DJs and remixers.

Three pieces of advice right up front— when selecting software:

1. Download a free demo for evaluation. Many manufacturers provide free demos on their web sites.

2. Read the system requirements carefully to determine if the software is compatible with your computer system's hardware and operating system (OS).

3. Try hard to speak with someone who is already using the software you are considering.

Technology progresses so quickly that sales people can't possibly keep up with everything, and even if they're good people, they're still trying to sell you something. Music technology magazines also have an inherent conflict of interest, as the same manufacturers who make the products reviewed in the magazine are also advertising clients.

Fig. 9.1. DJ RaeDawn's system incorporates CD turntables, a laptop, and MIDI bass pedals.

Some programs are cross-platform, meaning you can use them on both Windows and Macintosh computers, while others are platform and OS specific. If you're a PC loving, true blue nerd, you may be surprised by the high concentration of Mac applications here. Musicians discovered the Mac back in the pre-Windows days of MS-DOS, and music related users provide a sizeable constituency for the platform.

Software Categories

There are several categories of software of interest to DJs and remixers:

- Live DJ mixing and/or performing programs
- Looping and groove generating programs
- Musical Instrument Digital Interface (MIDI) sequencers (most include digital audio capabilities)
- Digital audio-based recording/mixing programs (most now include MIDI capabilities)
- Soft synths
- Soft samplers
- Software processors.

The last three on this list are often called "plug-ins," which are smaller computer applications that run **inside** of a "host" application. More about this later.

Programs like *Traktor* and *Live* have been designed specifically as mixing and performance environments for DJs and electronic musicians. While stand-alone applications such as Recycle are dedicated to the chopping and time manipulation of drum loops, Reason and Acid are working to establish themselves as "all-in-one" packages, by including beat making drum-looping capabilities, a MIDI sequencer, soft synths, and samplers.

ProTools built its reputation as a multi-track digital recorder and editor, and has added MIDI sequencing in recent years. Conversely, Logic, Digital Performer, Sonar, and Cubase initially built their strength around MIDI sequencing. MIDI sequencing is the recording and editing of musical events as data, played back on synths and samplers utilizing the MIDI protocol. This protocol can also be used to synchronize MIDI enabled drum machines, audio effects, lighting controllers, CD players, and other devices. MIDI sequencing is used by composers to write orchestral scores, and by electronic musicians to make all manner of groove-oriented tracks. All of the aforementioned sequencers also have integrated digital audio.

Soft synths and samplers, like Native Instrument's Kontakt and Reaktor, Arturia's classic analog synth emulations, and Logic's ES2 and EXS24 (bundled with Logic Pro), are designed to be accessed within sequencing/digital audio programs. The same is true for effects plug-ins such as reverbs, delays, EQs, and compressors, by companies like Waves, Ohm Force, Izotope, and Audio Ease.

As we check out these myriad tools, we'll also look at some of the features that may be most useful for DJing and remixing.

Tools for the Road

Many professional DJs are bringing their laptops to the gig, opening up a vast new landscape of possibilities. Rather than emanating from vinyl or CD, music coming from a laptop is usually stored on the computer's internal hard disk, often as MP3, MP4, AAC, Ogg Vorbis, or WAV files.

As hard disks become less expensive and processing power increases, it is becoming possible to use larger audio files, thus increasing fidelity. This is a welcome trend for DJs, as the lower-quality MP3s are not really high enough in quality to hold their own at high volume in front of a large audience.

Vinyl Emulation

Vinyl emulation software and hardware interfaces allow the user to physically manipulate the playback of digital music files on a computer using time code vinyl records on turntables, or time code CDs on CD turntables. This allows the user to scratch digital files as though they were on vinyl; also to cue, beat match, change tempo, beat juggle, or any other DJ or turntablist technique. This can also be accomplished using CD turntables or DJ CD players, using CDs to play back the time code rather than vinyl.

The output from the time code records is sent to an interface that connects to the computer (via USB, Firewire, or another protocol), where it is used to control audio files (moving them back and forth, faster or slower) through software that has been specifically designed or adapted to slave to the time code. The audio output from this software is routed to the same interface, where it is converted into analog sound and sent to a DJ mixer, allowing the DJ to interact with digital audio files stored on a computer's hard drive in much the same way as she would with music coming directly off of vinyl or CDs.

The overwhelming advantage to these systems is the space and weight of either the vinyl or the CDs that the DJ doesn't have to carry around anymore; a decent size hard disk can hold literally tens of thousands of songs. It allows DJs to bring their entire record collection to every gig, and to search for songs with the aid of a powerful computer, alphabetically, stylistically, by tempo, or from any number of tags that the DJ has placed on her music files. The potential danger is that of losing one's music collection once it becomes data, so multiple backup disks are a must. My rule of thumb is that data doesn't exist until it is stored in at least three places.

The potential downside of these systems is that we're potentially one step closer to becoming digital zombies, staring at our computer screens rather than interacting with the audience in front of us. Most of these systems have an "audio through" path for the turntables and CD players, allowing the DJ to switch to a regular record by flipping the phono/line switch on the DJ mixer. Scratch DJs who also mix will sometimes switch to regular records for their scratch feature, then mix through the computer using vinyl emulation.

Final Scratch

The first vinyl emulation system to find wide distribution was Final Scratch, which was invented by Mark-Jan Bastian, and premiered in 1998. Batian's Netherlands company, N2IT, made a deal with the DJ equipment manufacturer Stanton Magnetics to bring Final Scratch to market. Originally developed for BeOs, then Linux, the original version of Final Scratch (also dubbed FS) introduced the "ScratchAmp," a round USB/RCA interface about the size of a hockey puck (Figure 9.2).

Stanton began working with Native Instruments to offer a version of NI's DJ software, *Traktor FS*, with the product, and by Version 1.5, Final Scratch was Mac and Windows XP compatible, and included the ability to shift tempo and pitch independently of each other.

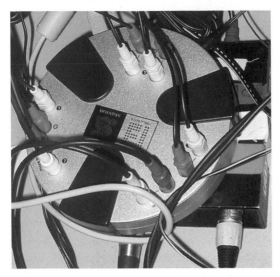

Fig. 9.3. Stanton's ScratchAmp 2 supports digital recording as well as play back.

Fig. 9.2. Final Scratch's original ScratchAmp has two stereo phono inputs and outputs, two stereo line outputs, and a USB port.

With Final Scratch Version 2, Stanton switched to a Firewire ScratchAmp, which features 24-bit/96 kHz digital quality playback and recording, MIDI in and out and an Audio Stream Input/Output (ASIO) driver. Version 2 is also compatible with NI's *Traktor DJ Studio* (starting at Version 2.6), which is more powerful than *Traktor FS*.

Native Instruments and Stanton scaled back their relationship in late 2005, with NI making deals with other hardware manufacturers and Stanton releasing *Final Scratch Open*, which consists of the ScratchAmp 2 and drivers, but no dedicated software application.

Serato Scratch Live

In 2004, the New Zealand-based company, Serato Audio Research, teamed up with Rane, an audio hardware manufacturer known for their DJ mixers, to bring *Serato Scratch Live* to market. Recently donning the nickname "SSL," Serato Scratch Live, should not be confused with the *other* SSL, Solid State Logic, the high-end console manufacturer. Serato's SSL is similar in concept to Final Scratch, with time code records and CDs, and a USB/RCA interface which includes a mic input for recording. SSL's main calling card is its dedicated software, which plays up the vinyl interface and seeks to make mixing on a computer into a visually rich and intuitive experience.

The waveforms in the Track Overview, Main Waveform, and Beat Matching displays are colored according to the spectrum of the sound: red representing low-frequency bass sounds, green representing mid-frequency sounds, and blue representing high-frequency treble sounds. This usually translates into kick drum transients being red, and snare drums being green or blue.

In SSL's Tempo Matching display, *Scratch Live* detects the beats within the track, and places a row of orange peaks (for the track on the left side) above a row of blue peaks (for the track on the right side) in the Tempo Matching display area. When the two tracks are matched to the same tempo, the peaks will line up.

Compared to programs like *Traktor DJ Studio*, *Serato Scratch Live* has a limited feature set. As of this writing, the program is just now incorporating a basic pitch lock feature, which seems odd since Serato is known for its ProTools plug-in, *Pitch 'n Time*. Still, SSL's simplicity and stability has won it a large group of dedicated users. Rane also makes a mixer with the computer/audio interface built-in, making set-up simpler and adding dedicated hardware control for some of the softwares' functions.

Fig. 9.4. Serato Scratch Live gives the user four different views of the music's waveform, the Tempo Matching display across the top, the Track Overview display along the sides, the Main Waveform display (the largest waveforms giving current position), and the Beat Matching display in the center.

Fig. 9.5. Numark's Virtual Vinyl running Cue as its software interface includes the ability to mix video files as well as audio files.

Entering the Fray

Currently, there is an explosion in the number of companies bringing vinyl emulation systems to market.

Numarks' system, *Virtual Vinyl*, ships with a program called *Cue*, which includes the ability to mix and scratch video files, create beat-matched loops, and incorporate a built-in effects and Virtual Studio Technology (VST) plug-ins (more on these later).

The Torq Conectiv Vinyl/CD Pack from M-Audio, Ms Pinky's Interdimensional Wrecked* System (IWS), and new hardware from Mackie, Allen and Heath, and others make this a rapidly growing and changing field.

There is also a trend toward doing away with the turntables and CD players altogether, and replacing them with a dedicated control surface. Entries into this market from Hercules, M-Audio, PCDJ, Vestax, and Numark show that the people who think this stuff up think that this is the future of DJing.

Ableton's Live

Live is an audio sequencer designed to be flexible in a live performance environment. Ableton contends that Live "transforms rigid audio into elastic matter," and the program goes pretty far to justify this claim.

Fig. 9.6. *Live* has been designed for recording, arranging, experimenting, and improvizing.

Techniques previously confined to the studio, such as stutter edits, reversing samples, and beat mashing, lead many feel as though Live is the first sequencer/DAW you can perform and improvize with as though it were a musical instrument. Live lets you rearrange beats on the fly, incorporating a flexible and sophisticated triggering method which lets you combine any number of samples with each other.

You can import samples of any tempo into Live and adjust their tempo to match the track you are already playing, in real time without changing pitch. You can also change the pitch of an audio file independent of tempo.

The automatic beat-matching and real-time quantization features mean that even long pieces with tempo variations can play back in sync.

During playback you can drop in samples on the fly without stopping the sequencer, a handy feature for live performance. You can also go back after the fact and do more detailed editing in the studio.

Live is both Mac and PC compatible and can be integrated with ProTools or your favorite sequencer as long as they support ReWire (more on this later).

Traktor

A similarly flexible real-time program is Native Instruments' *Traktor DJ Studio* (for Mac and PC), which was developed in cooperation with professional DJs. It includes a track database with a very fast search function, a DJ mixer, on the spot beat matching, up to 10 tempo-precise loops and cue points which you can set on the fly, mix recording, overdubbing, and

Fig. 9.7. Traktor takes the DJ environment (complete with a crossfader) into the virtual world.

fast mix export capabilities. You can even automatically import iTunes playlists and libraries (however you can't use DRM-encoded tracks purchased from the iTunes site).

Traktor interfaces directly with Beatport, a digital downloading network designed for DJs and dance music, allowing you to purchase new tracks without leaving the program.

Native Instruments has been busy lining up partnerships with various DJ companies, and even has developed its own time code record and CD interface system dubbed "Traktor Scratch." A partnership between Native Instruments and the hardware manufacturer Allen and Heath has resulted in a mixer with the digital interface guts built-in.

All-in-One Programs

Based on the same concept as the all-inclusive production centers, software developers started creating software programs that excel at making beats, sequencing, and slicing and dicing audio, just like their hardware counterparts. The programs *Acid* by Sonic Foundry and *Reason* by Propellerhead are like virtual versions of the Roland TB-303, TR-808, and Groovebox series, and in many ways even resemble them in their design and layout concept.

Acid Pro

An intuitive paint-and-play interface along with a continuously expanding host of features and beats has made Acid Pro a tool of choice for those wanting to quickly create groove-oriented, loop-based music (Fig. 9.8).

Acid makes it so easy to "make beats," that many people who principally use ProTools and other programs will use Acid to get a groove happening, then import that groove into another program as an audio file. Acid Pro saves to a variety of formats such as WAV, WMA, RM, AVI, and MP3.

Acid's strength lies in real-time pitch and tempo matching. This means you can take any loop that has been "Acidized" (meaning it contains pitch and tempo information) and scale it to play in whatever tempo and key you are currently working in. In Acid, you can combine loops with original tempos of 73, 89, 122, and 155 bpm beats per minute) and play them all together perfectly at 140 bpm.

The Beatmapper, with its ability to detect an existing song's tempo when imported, is invaluable to remixers. With the help of the Chopper you can cut and slice samples to

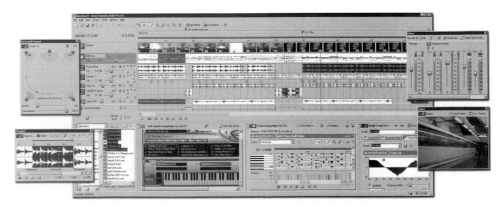

Fig. 9.8. Acid Pro with video, surround panning, and sequencer windows open.

create drum fills, stutters, and DJ style effects. It's even possible to import a variety of video formats and use Acid to score for picture in 5.1 surround sound, provided you've got the right interface and monitoring environment.

Acid features a MIDI sequencer with the ability to use DLS soft synths and VST instruments (VSTi). We'll examine these software synthesizers and samplers in more detail a little later.

Acid also comes with scores of effects plug-ins, which you can apply to tracks or busses in real time (for interesting morphing EQs and filter sweeps), or string together to create your own effects chains.

Acid is PC-only software. There's no Mac version, but I know more than one Mac user who bought a PC just to run Acid.

Reason

The concept of making a computer program actually look like the gear it's replacing helped Reason make quite a splash when it was first introduced. For those of us old school enough to have learned on the hardware, there is something reassuring about the familiarity of Reason's GUI (Graphical User Interface, pronounced "Gooey") (Figure 9.9).

In addition to its practical interface, Reason is capable of delivering serious power to those interested in cranking out pulsing, groove-oriented tracks. Made by the Swedish company *Propellerhead*, Reason is a production tool and music workstation that consists of several modules and a sequencer, all in software form. *Propellerhead* looked at the typical rig of musicians, DJs, producers, and remixers who use sequencers, drum machines, synthesizers, samplers and mixers, and made software versions of each component. Reason is available both for Mac and PC.

Fig. 9.9. Reason's interface resembles a rack of audio and MIDI hardware. You can even look behind the rack to check (and change) connections with virtual patch cables.

In Reason's virtual rack, a *module* can be an instrument (like a sampler or a synth), an effect processor (like a reverb or flanger), a mixer, a matrix, or a patchbay. On screen, modules look like rack versions of popular devices, complete with buttons, switches, faders, and LEDs. You can even look behind your rack and hook things up with virtual cables. Modules can be connected to each other in conventional and unconventional (experimental) ways.

Sure, you might be able to find some better sounding soft synths, a more capable soft sampler, and better-equipped sequencers on the market, but for a lot less money Reason gives it all to you in one package that was designed to work together. Reason is also an excellent learning tool. You can learn a lot about signal flow, sampling, synthesis, and basic mixing techniques all in one program.

ReWire

ReWire, a software protocol jointly developed by Propellerhead and Steinberg, facilitates audio and MIDI interconnectivity between an ever-growing number of applications. This makes Reason an even more complete package, as it allows Reason to interface with ProTools, Logic, Digital Performer, Cue Base, and other digital audio equipped sequencers that contain more extensive audio features. ReWire lets you access Reason's soft synths from a host sequencer, and trigger them just like outboard MIDI synth modules.

ReWire is also being used by programs such as Ableton Live and Melodyne to integrate with other applications; in fact most modern sequencers support ReWire and work as ReWire hosts (or "mixer applications"), sending MIDI messages to the ReWire client (or "synth application"), and accepting audio from the synth application. A ReWire system can contain only one host, but multiple client applications can coexist. Unfortunately, as of this writing, Ableton Live, Melodyne, Reason, and FL Studio are among the only sequencers that have the ability to work both as hosts and clients, as most sequencers have been designed by their manufacturers to work only as hosts. This makes it impossible, for instance, to use soft synths from Logic within a ProTools session through ReWire.

ReWire continues to gain ground as an industry standard, and many soft synths and soft samplers, like GigaStudio, Obsession, Retro AS-1, and Unity DS-1, will slave to mixer applications through ReWire.

REX Files and Recycle

Reason contains the "Dr. Rex" module, which triggers and manipulates REX files, which we'll explore next. Instead of working with a regular audio file of a drum loop, think of accessing each hit within that loop as a separate slice—that's exactly what REX files are.

You can download REX files or you can create your own in Propellerhead's Recycle. Once a groove is loaded into Recycle, it's analyzed and sliced up into rhythmic components; separating kick, tom, high-hat, and snare hits. The groove's tempo and pitch can now be changed without one affecting the other (Figure 9.10).

The newly created loop can be saved as a new audio file in AIFF format, transmitted to a sampler, or it can be saved as a REX file and imported into programs that support this format (such as Logic, Cubase SX, or Reason). Recycle can also create a MIDI file based on the timing of this groove. Once the MIDI file is imported into a sequencer, it will trigger the slices you transmitted to the sampler and play back the entire groove, giving you a

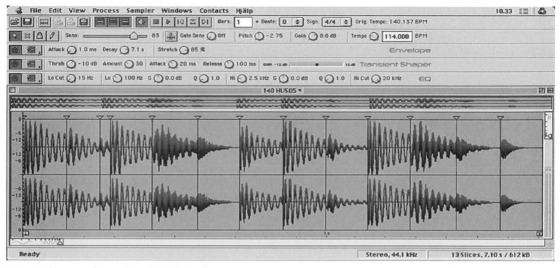

Fig. 9.10. A glimpse of Recycle's slicing interface.

great deal of editing control over the groove's components. You can also replace individual sounds, like kicks and snares, as you see fit.

Fruityloops/FL Studio

FL Studio began life as an easy to use application called *Fruityloops*, which was distributed for a time by Cakewalk. Originally a simple pattern-based sequencer, it got the job done and

Fig. 9.11. The FL Studio began life as the simpler software package known as Fruityloops.

it sounded surprisingly good. Quite a few dance music producers and DJs used Fruityloops early on, although they were not likely to admit it in interviews.

FL Studio was introduced for higher-end users, and eventually supplanted Fruityloops, although there is still an FS Studio Fruityloops Edition available. All FL Studio versions (except for FL Studio Express) are VSTi, DXi, Buzz, and ReWire compatible, and support WAV, MP3, and OGG audio formats.

FL Studio Producer Edition and the top of the line XXL version include extensive audio recording and editing capabilities, and come with an array of internal effects, a 64-channel mixer, and a cool Wave editor called *Edison*, which is a recording and editing tool that includes spectral analysis and a convolution reverb. All FL Studio versions run on Windows machines only (Figure 9.11).

ProTools

The most ubiquitous professional digital audio recording/mixing program in the USA is ProTools by Digidesign. One of the first multi-track digital audio applications to catch on at the professional level, ProTools' streamlined intuitive interface has made it the industry standard for recording, non-linear, non-destructive editing, and mixing.

While Digidesign used to offer a limited version of their software (ProTools Free) as a demo, the company stopped updating this version years ago. ProTools is a hardware/software system, therefore, you must be using Digidesign (or M-Audio) hardware in order to run ProTools. Digidesign purchased M-Audio a few years ago, and as of this writing, the least expensive way to start using ProTools is to purchase *ProTools M-Powered* and run it on one of M-Audio's inexpensive audio interfaces.

The next step up is to purchase a ProTools LE system with one of Digidesign's ProTools LE audio interfaces. LE and M-Powered versions of ProTools are powerful, although they are limited compared with ProTools HD.

For Digidesign's higher-end users, there's ProTools HD (or High Definition). HD systems are capable of recording at sample rates of up to 192 kHz, and utilize PCI or PCI express cards as well as HD interfaces with excellent DA/AD converters supporting up to 96 channels of I/O (inputs and outputs), supporting play back of 128 tracks of audio.

Working with Beats and Loops in ProTools

While not originally developed as a sequencer or a beat-making application, ProTools now offers many functions that come in handy when you're remixing and working on loops and beats. When selecting the "grid" mode, you can snap MIDI notes and audio regions to a user-definable time grid. This allows you to accurately align loops or single instruments (like a kick or a snare) on the beat.

Figure 9.12 shows how the high-hat, kick, and snare are aligned with the drum loop (in purple) by snapping to the 16th note subdivisions marked by the light blue lines. Notice how the MIDI kick below the loop is also perfectly lined up with the red audio kick and the loop. Working in grid mode with the grid representing bars and beats makes creating and modifying grooves much easier than working with a timeline representing minutes and seconds.

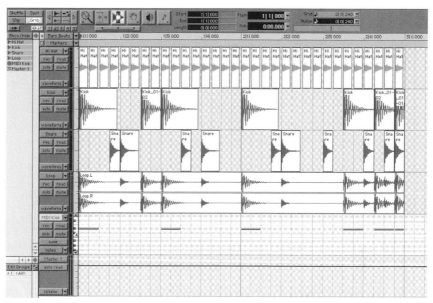

Fig. 9.12. A simple and tight groove in grid mode, combining an audio loop with a MIDI kick and audio samples for high-hat, snare, and kick drum.

Plug-ins

Plug-ins are tools that can do all sorts of useful things: tune flat vocals, replace lousy sounding drums, emulate tape compression, or add practically any sort of audio effect ever dreamed up. While ProTools ships with some basic *Digirack* plug-ins (reverb, compression, EQ), you can increase the power of the system immensely with third-party plug-ins. There are dozens of amazing plug-ins out there for ProTools, and most plug-in manufacturers also make versions for other programs, like Logic and Digital Performer.

When talking about plug-ins we need to differentiate between *real-time* and *audio suite* plug-ins. Real-time plug-ins (RTAS in the ProTools LE environment) are inserted into an audio channel and affect your track in real time as it plays back. Real-time plug-in processing and automation in the LE environment is *native*, meaning it is dependent on your computer's processor (host CPU); the faster your computer and the more RAM you have, the more real-time plug-ins you can use at once. The real-time plug-ins within the HD system are *TDM plug-ins*. A dedicated DSP chip on a PCI card powers TDM plug-ins, consequently sparing your computer's processor from doing the math.

In contrast, *audio suite* plug-ins are file based: when applied to an audio region, you wait while they create a new audio file with the selected effect, replacing the original audio region. This usually only takes a few seconds, depending on the size of the file being processed, the complexity of the processing, and the speed of your CPU. Consequently, audio suite plug-ins don't exhaust your computer's processor while mixing. However, if you want to tweak a few parameters on an audio suite plug-in a little later, you have to wait for it to write a new file. If you think you might want to go back to the unaffected audio file, you need to make a copy (with a name you'll remember) before you write over the original sound file.

Fig. 9.13. Sound Replacer lets you keep a performance while replacing less-than-stellar sounds with new samples.

Sound Replacer

When creating tracks for the dance floor it can be effective to layer two or three kick samples; one for the attack or thwack, one for the deep low end, and one for the body or tone of the kick. A useful plug-in for replacing kick drum (and snare) tracks is Digidesign's *Sound Replacer*, which allows you to replace or blend up to three different samples within a single audio selection, creating a beat with a new texture. Sound Replacer has three separate threshold levels, each corresponding to a different sample, and various cross-fade, peak alignments, dynamics, and mixing functions. The cross-fade function molds the new sample into the audio region smoothly, maintaining the original track ambience. Sound Replacer works with both LE and HD systems as an audio suite plug-in (Figure 9.13).

Synchronic

Synchronic is a Digidesign RTAS plug-in aimed at DJs and remixers wanting to create rhythmic and sonic variations in the manipulation of audio loops. Synchronic can be used to quickly create rhythmic modifications and in-tempo effects with individual beats and beat subdivisions within a loop (Figure 9.14).

Some other popular third-party plug-ins useful to DJs and remixers include Serato's *Pitch 'n Time*, which lets you do radical tempo and pitch adjustments independently of each other, Auto-Tune by Antares for tuning up vocals and other monophonic tracks, and Melodyne which is another intonation plug-in that adds useful time correction tools. The WAVES plug-in bundles, which include high-quality reverbs, compressors, and EQs, are also popular, as is the convolution reverb program Ultaverb. Most popular plug-ins can be purchased in either native or TDM versions.

As of this writing, ProTools TDM software features some additional functions, such as SMPTE time code compatibility (useful when syncing music to external video), *Replace Region*, and multi-track *Beat Detective*. The *TC/E Trim Tool* functions, originally only available on TDM versions, is now also available in ProTools LE, as is a limited version of Beat Detective. Digidesign also comes up with some creative ways of packaging some of these

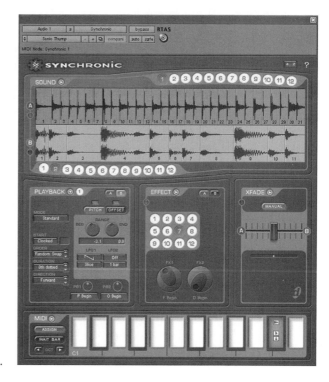

Fig. 9.14. The Synchronic plug-in by Digidesign.

features into products like the *Music Production Toolkit*, which adds features to an LE system—for a price.

Replace Region

Replace Region allows you to replace multiple instances of an audio region with another sound file. Let's say you don't like a particular snare sound and are wondering what it would sound like if you used a different sample instead. Select the snare you want to replace and then Command-drag (Macintosh) or Control-drag (Windows) the replacement region to the selected region and bam—you can now listen to the same performance with a new snare sound. It's not as smooth as the previously-discussed Sound Replacer plug-in because samples are not being blended together, but replaced entirely. Of course, this may be exactly what you're after.

Beat Detective

Beat Detective is a rhythm analysis and correction tool that extracts tempo information from a loop by detecting its peak transients and generating bar and beat markers. Based on this information, another loop's tempo can then be conformed to the just generated session's tempo map. This tool can also simply be used to cut up a performance into separate bits (Figure 9.15).

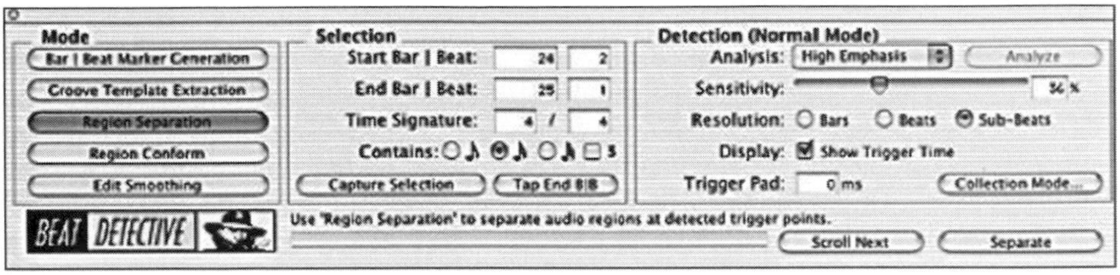

Fig. 9.15. Beat Detective adds welcome remixing functionality to ProTools.

TC/E Trim Tool

The Time Trimmer will match an audio region to the length of another region or a tempo grid. This tool uses the Digidesign's Time Compression/Expansion plug-in in order to create a new audio file. If you use Time Trimmer in grid mode, you can import a drum loop of a different tempo than your current session, snap it to the beat and then quickly time stretch it. You have to be careful though, because too much time compression will start affecting the audio quality of your sample.

If you are doing extreme time-stretching, you may be better off working with Serato's Pitch 'n Time plug-in, which uses a better sounding algorithm. Note: in the ProTools preferences you can set the Trimmer Tool to use the algorithms of third-party time compression/expansion tools you may own, like Waves' Time Shifter, or Serato's Pitch 'n Time.

Sequencers

While ProTools is currently one of the industry's leading audio programs, its MIDI sequencer is not as advanced as some of the competition. This is not to imply that ProTools' MIDI sequencer is not powerful, but for those who are MIDI sequencer aficionados, it has tended to be behind the curve on features. Soft synths came late to ProTools, and other features, like extracting written notation from MIDI files have been late in coming as well. Most of the important features found in MIDI sequencers have been incorporated into ProTools, and more users are using ProTools as a primary sequencing environment every year. However, many composers and remixers who are heavily involved in MIDI sequencing choose to use another program for sequencing, even if they wind up mixing in ProTools.

Bear in mind that there isn't one sequencer that will please everybody; choosing a sequencer is ultimately a very personal decision.

With the exception of M-Audio versions (ProTools M-Powered), Digidesign sotware only works with Digidesign hardware. This means that you can run Digital Performer with a Digi 002, but you won't be able to run ProTools using a MOTU 828mkII audio interface.

Most other programs will talk to a MOTU UltraLite, an Aardvark PCI card, or an RME PCI card—just to mention a few. It's best to double-check an application's requirements first in order to be absolutely sure that you have a compatible software/hardware situation. I personally have a rule that I will not buy a software/hardware combination unless I talk to someone who is already running that configuration with good results. I've just been burned too many times.

The sequencers we'll be discussing offer most of the same core features, implemented in different ways. In addition, their GUIs may look quite different from one another. It's a good idea to play around with as many programs as possible in order to find what's really best for you. While there are dozens of sequencers on the market, most professionals in the US gravitate toward Steinberg's Cubase, Logic Audio, Cakewalk's Sonar, or MOTU's Digital Performer.

 Logic

Logic is a MIDI sequencer with extensive digital audio capabilities popular among professional DJs, remixers, composers, and electronic musicians. Unlike ProTools, Logic will talk to most soundcards and audio interfaces.

The development of Logic Audio for the PC was discontinued in Fall of 2002 when Apple Computer bought Emagic. If you're a Mac user, Logic Audio is an extremely powerful program you should consider. Some people never get past Logic's rather steep learning curve, but once you spend some time with it, you'll be rewarded by incredible flexibility. There is no other sequencing program which lets you customize and personalize your work environment as extensively. Logic gives you the ability to configure 90 layout screens and to recall them with the help of the numeric keypad on your keyboard (Figure 9.16, 9.17).

Some users claim that Logic Audio is also one of the "better sounding" programs in terms of digital sound quality. Logic has also made a name for itself because of its zooming feature, which lets you get down to the sample level, allowing for highly sample accurate editing. Users like BT make extensive use of this feature.

Fig. 9.16. Logic Audio's layout is extremely customizable.

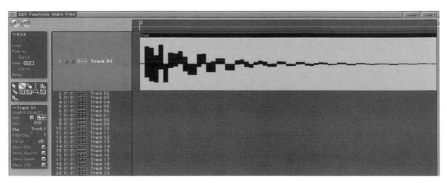

Fig. 9.17. When you zoom down to the smallest resolution in Logic, the waveform resembles a series of rectangles, each being a visual representation of a single sample bit.

Many remixers love using Logic because of how well MIDI and audio are integrated in this program. Logic also comes bundled with over 50 high-quality plug-ins, ranging from compressors, EQs, reverbs, and delays (including Space Designer, a convolution reverb), to filters and numerous special effects.

Cubase

Cubase by Steinberg has been a popular sequencing program since its introduction in 1989. Steinberg is the pioneer of VST, with Cubase VST becoming the first native software to

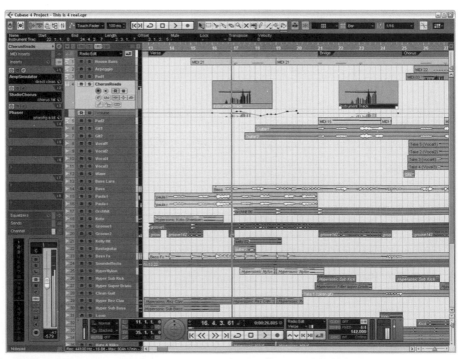

Fig. 9.18. Cubase offers seamless integration of VSTi, effects, and digital audio.

incorporate this real-time studio environment including EQs, effects, mixing, and automation back in 1996. The Cubase SX line for Mac and PC replaced Cubase VST in 2002.

With Cubase, you get a lot of nice VSTi, effects, and soft synths, like the *lm-7 Drum Machine*, the *a1 Analog Synth*, and the *vb1 Virtual Bass*. Loops can be sliced up in a manner reminiscent of Recycle. *Arpache 5* is a MIDI plug-in arpeggiator with customizable presets—very useful for creating pulsating electronic dance music. *Step Designer* is a great step sequencer where you can draw notes with the mouse (very similar to Reason's Matrix), add controllers, and even randomize the notes. You can then save your patterns as presets and recall them in other projects.

Of interest to DJs are Steinberg's Xphraze phrase synthesizer, HALion VST sampler, and Groove Agent virtual drummer, all of which integrate with Cubase or any other VST compatible sequencer (Figure 9.18).

Sonar

Sonar by Cakewalk is a PC-only sequencer that has drawn a lot of attention for a lot of years. Its grid and pattern-based drum editing and custom drum mapping, together with soft synth integration, DXi, and ReWire support make it a handy package. You can export ACID-format WAV files for use in other projects and applications. Sonar currently has many editions available, some of which ship with two 64-bit, fully automatible DirectX 8 mastering effects, and an advanced DXi soft synth drum sampler. In addition, the fully automatable, 64-bit Timeworks EQ and the 30-band spectrum analyzer and phase meter are excellent tools for mastering. Remixers will be interested in the audio loop tools, which have features similar to programs like Sonic Foundry's Acid and Recycle in terms of beat detection, tempo/pitch changing and beat matching. They'll also be interested in Rapture, Cakewalk's new beat creation tool (Figure 9.19).

Fig. 9.19. Sonar by Cakewalk has many features friendly to DJs and remixers.

Digital Performer

Digital Performer by MOTU, or *Mark Of The Unicorn*, runs only on the Mac and has evolved into quite a nice sequencer over the past few years. One handy remixing feature in DP (as it is referred to by most users) is the *Adjust Soundbites to Sequence Tempo* feature. If you have a MIDI sequence that you want to enhance with a drum loop, but the tempo of the loop is different than your sequence tempo, DP will adapt the loop to the sequence tempo by time stretching it. The exact opposite is also possible with the *Adjust Sequence to Soundbite Tempo*.

Another valuable tool in DP is the *Drum Editor*, which allows you to visually arrange MIDI triggered drum components within a grid-like window, quantize them, adjust their velocity curves, and rearrange them as you please (Figure 9.20).

MOTU has been aggressively going after the hardware market lately as well. All MOTU hardware ships with ASIO drivers for compatibility with most popular sequencers and OS, Windows Vista included. ASIO stands for *Audio Stream Input/Output* and is Steinberg's low-latency/high-performance standard protocol.

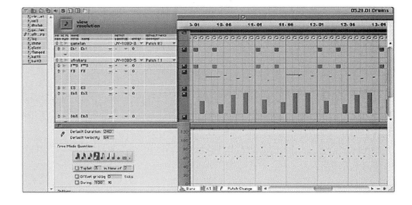

Fig. 9.20. Digital Performer's Drum Editor window.

VST and Audio Units

VSTi is short for Virtual Studio Technology Instrument. First developed and launched by Steinberg in 1996, they are what their name implies, "virtual instruments," like software (or "soft") synths, samplers, drum machines, etc. Cubase VST first introduced *Neon*, the first in a series of virtual instruments for the Steinberg product line. This simple soft synth is a virtual two-oscillator synthesizer with a resonant low-pass filter (Figures 9.21 and 9.22).

Most sequencers are now VST compatible and will allow you to access these instruments within the program as channel inserts. You can trigger and control their parameters via MIDI, but at the same time you can treat their output just like audio and apply effects to them.

VST effects plug-ins emulate hardware effects processors and are accessible like other plug-ins in your sequencer. A downside to VSTi and plug-ins is the amount of processing power they often require, making a fast CPU and lots of RAM even more important.

DXi is Cakewalk's answer to Steinberg's VSTi, and is integrated in Cakewalk's Sonar line. While recent versions of Sonar are also compatible with VSTi, you can bridge some

Fig. 9.21. Neon is the pioneer of all VSTi.

Fig. 9.22. The EXS24 plug-in by Emagic is a pioneering soft sampler.

incompatibility with the use of "Wrappers," which fool a host application into using a format they don't specifically support.

On the Mac side, Apple has developed a system-level plug-in architecture called Audio Units, which is similar to VST. Most Mac applications (except for ProTools) make use of Audio Units rather than VSTi. ProTools uses its own Audio Suite and RTAS (real-time audio suite) format, as well as TDM (Time Domain Multiplex) for ProTools HD systems.

Most widely distributed plug-ins ship in multiple formats these days, but it's important to make sure your system is compatible before you buy. For more information on soft synths, samplers, and plug-ins is check out http://www.kvraudio.com. Here you can find compatibility charts and the latest tips and tricks.

Conclusion

This vast amount of available technology opens a wealth of possibilities, but can also be overwhelming. Many of these products do similar things; the best strategy is to choose a product and learn it really well rather than always being seduced by having the new thing.

Your best defense in making sure you choose the right tools for your needs is self-education and due diligence in finding people to speak to about a product who don't have a vested interest in you spending your money. The Internet can be a valuable tool in this quest, but don't believe everything you read.

Choose your gear wisely by clearly setting your goals and envisioning what you want to achieve. Remember, these are only tools; the true creative source is you.

10

Drum Machines and Groove Boxes

Drum machines have been part of the DJ and remixer's arsenal of gadgets since shortly after they appeared on the scene.

The precursor to the drum machine was the *beat box* or *rhythm machine*, which played back preset rhythms. These became a staple on electric organs, and had a pretty high cheese factor (which can occasionally be desirable, as rhythm machines have made cameo appearances on some cool records). User-programmable drum machines first started appearing around 1980, finally allowing musicians to create their own grooves.

Simply put, a drum machine combines digitally sampled percussion sounds and a sequencer. As we saw in Chapter 9, a sequencer is a device that allows you to record and manipulate notes, then play those notes back in order (or in sequence). These notes are not recorded as audio, but as performance events into a computer, allowing the exact placement, velocity, and the length of those notes to be easily manipulated.

Digital sampling, the practice of making short digital recordings of live instruments, then using a keyboard, drum pad, or other controller to play back these sounds, made it possible to put any sound under sequencer control.

In the early 1980s, sequencers were most commonly incorporated into drum machines and Musical Instrument Digital Interface (MIDI) keyboards. Machines like the AKAI MPC series added the ability to sample your own sounds in the late 1980s.

Sequencers later became the core of most computer-based MIDI workstations in form of sequencing software such as Master Tracks, Logic, or Performer. In the 1990s, digital audio was added to software sequencers making them hybrid tools where sequenced notes and recorded audio can be mixed together and manipulated within the same application. For a thorough discussion of sequencers, see Chapter 9.

While a powerful laptop running the latest software applications can do more that even the most up-to-date drum machine/sampler/sequencer, the robustness, sound, and feel of the hardware machine and the specific tactile nature of working on these machines had kept them popular.

Welcome to the Machine

The world's first programmable, sample-based drum machine, the LM-1, was designed by Roger Linn in 1979. As a musician, Linn was interested in developing a drum machine that was customizable and capable of producing its own patterns beyond the common preset "samba" and "mambo" beats of the rhythm machines. He used samples of acoustic drum sounds played by LA session drummer Art Wood. The LM-1 had 18 drum sounds, sampled at 28 kHz, which consisted of kick, snare, rimshot, high-hat, toms, cabasa, tambourine, high and low congas, cowbell, and claps. Linn was not able to include cymbals because sampling sounds with longer decays would have been too costly at the time. This drum machine was capable of playing 100 patterns in real or step time and chain them together to create up to eight songs. Other innovations of the time were the swing and quantize functions (Figure 10.1).

Despite the fact that only 500 units were manufactured, the LM-1 revolutionized the popular music of the 1980 and can be heard on classic recordings by Phil Collins, the Thompson Twins, Stevie Wonder, Depeche Mode, Jean-Michel Jarre, The Art of Noise, and Prince.

The LM-1 was followed by the LinnDrum, and later the Linn9000, which included cymbals, a higher sampling rate of 35 kHz, additional controls and live drum trigger inputs.

Le Chanson de Acid

Ikutaro Kakehashi founded the Roland Corporation in 1972, naming his company after a French poem, Le Chanson de Roland. Tadao Kikumoto was the man behind the Roland TB-303 (Transistor Bass) and TR-606 Drumatix (Transistor Rhythm), both created in 1982. Although intended to emulate a real bass player and drummer, these two units didn't sound much like live musicians. Programming proved to be time-consuming and cumbersome as well, leading to a halt in their production shortly after their release. However, their uniquely warm, punchy, and dirty sound quality drew the attention of DJs in the late 1980s who gave them a new lease on life.

Phuture (DJ Pierre, Spanky, and DJ Herbert J) re-introduced these boxes to the dance and electronic music world as the distinct sound of a genre called *acid*, a term which has become a popular modifier (as in acid house, acid techno, etc.). *Acid* in the name of a genre sometimes implies the presence of the TB-303 bass sound.

The Roland TR-808 and the TR-909's hard kicks and ubiquitous claps can be heard on countless 1980s pop recordings and electronic dance tracks through to the present.

Fig. 10.1. The LM-I LinnDrum, circa 1979.

Alesis SR-16: The Long Run

The Alesis SR-16 is a compact, inexpensive drum machine that has remained popular since 1990. It's more of a "real" sounding drum machine, featuring 233 realistic, natural drum sounds, many sampled both dry and with reverb. Velocity-sensitive pad buttons enable drum sounds to change tonal content as they're played harder—not unlike acoustic percussion instruments (Figure 10.2).

The SR-16's sequencer is designed to only control onboard sounds, and includes some preset patterns that were played in by live drummers. MIDI Clock/Song Position Pointer allows the SR-16 to sync with other MIDI devices, and an external MIDI sequencer or controller may address its onboard sounds.

Groove Boxes and Production Centers

The Linn9000 was the first combination of a drum machine and a multi-track MIDI sequencer, meaning that its sequencer could control sounds from external MIDI devices. The Roland MC-505 groove box is an example of a sequencer/sound module with retro-styled dance sounds, featuring the classic TB-303, TR-808, and TR-909 sounds.

Units housing a built-in drum machine with a sequencer, sampler, onboard sounds, effects, and a variety of controls soon became the new trend and the term "production center" was born.

Roger Linn began collaborating with the Japanese company Akai in 1988 to design sampling and sequencing drum systems, which became (and remain) a staple in Hip-hop, rap and dance music production.

As it turned out, sampling drum machines were perfectly suited for the task of extending (looping) breaks off of records, and beefing those breaks up with additional kick drum and snare sounds. Since extending breaks is one of the core techniques in Hip-hop, dating

Fig. 10.2. The Alesis SR-16 made its way into many records because of its convincing sounds.

Fig. 10.3. The popular Akai MPC2000XL can read sample CDs of many formats through its SCSI port.

all the way back to DJs Kool Herc and Grandmaster Flash, the MPC was instantly adopted by Hip-hop producers, many of whom were also DJs.

In 1994, Akai released the Akai MPC3000 MIDI Production Center, which featured stereo sampling and onboard effects, as well as filters and timing that gave it a reputation for great sound and a rock solid groove that many Hip-hop producers and DJs felt was more solid than computer-based software sequencers.

The popular MPC2000XL offered new sample editing capabilities like time stretching, re-sampling, and a trim mode that let you slice a sample into 16 zones and assign each zone to a separate pad. The big, touch sensitive pads have always been an attraction of the MPC series (Figure 10.3).

The MPC2500 incorporated the features of the 2000XL, like eight assignable outputs, and added the option of housing an internal hard drive.

The flagship incarnation of the MPC is the MPC4000, a studio in a box supporting high definition 24-bit, 96 k digital recording, and even includes a phono pre-amp for direct connection of a turntable.

Korg Electribes

Korg's *Electribe* series is a collection of music production centers geared toward electronic dance musicians, DJs, and remixers. The warmth and thickness of the Electribe's sounds are meant to be reminiscent of retro analog synths. The ES-1 Electribe Rhythm Production Sampler, for instance, is a 12-track 64-step sequencer, sampler, and drum machine. You can use pitch, filter, and effects knobs to drastically alter samples. Its recycle-like *time slice* function lets you cut phrases into pieces allowing you to create your own patterns from simple phrases (more on recycle in Chapter 15) (Figure 10.4).

The more upscale Electribe EMX-1 and the ESX-1 both feature twin vacuum tubes for analog warmth. You can achieve gated patterns by feeding external audio such as sustained pads from another synth into the two audio inputs, and process it through the internal resonant low pass filter and effects. That's a handy way of creating the stutter and filter

Fig.10.4. Korg's ES 1—Electribe S Rhythm Production Sampler.

Fig. 10.5. The ESX-1 Electribe features vacuum tubes on the output stage, a ribbon, and slider control.

sweep effects popular in many dance and electronic tracks. The tempo of sampled loops can be adjusted without affecting their pitch. *Motion sequencing* is a cool feature that records your real time knob movements and plays them back as part of your programmed pattern. Electribe's will sync to external sequencers and other MIDI devices through MIDI clock (Figure 10.5).

11

Mixers and Effects

Not long ago, there weren't many choices when it came to audio mixers. Radio stations and recording studios had to hire engineers to custom build their mixing boards into the early 1960s.

In the 1970s Richard Long, Alex Rosner, and Louis Bozak took major steps forward in the development of the modern DJ mixer, with input from David Mancuso and others. The Bozak CMA-10-2D and CMA-10-2DL were two of the first commercially-available DJ mixers. With their big rotary pots and switches, the Bozak mixers concentrated on sonic quality and durability. Since Mancuso preferred to switch between songs with his hands on two rotary pots, there was no crossfader. Later, Bozak made a crossfader available in a separate unit that interfaced with the mixer.

DJ mixers have changed significantly since then, adapting to the evolving needs of the DJ. There are still club mixers with rotary pots that resemble the original Bozak, but there is also a huge variety of commercially available mixers that cater to club DJs who prefer faders over pots, and mixers that cater specifically to scratch DJs.

Function

Mixers take audio from a variety of sources, combine it all together, then send the mixed signal on to a PA system, recorder, radio transmitter, or some other destination. A professional mixer's fidelity must be impeccable, the monitoring flexible, the equalization (EQ) powerful, and musical. The metering should be accurate and logical, and there should be an easy way to interface outboard effects.

The unique task of a DJ mixer is to make it possible for the DJ to segue between sources in a seamless and elegant fashion, live. Scratch mixers also need to have precise controls that feel good enough to be played like a musical instrument, and can be contoured to an individual DJs style.

Professional DJ mixers also need to contain high-quality phono pre-amps, which contain a special EQ curve developed by the Recording Industry Association of America (RIAA). With the exception of modern turntables like the Numark TTX1, which have a phono pre-amp built in, all turntables must go through the RIAA phono EQ before being amplified, or the sound will be tinny and unusable.

Fig. 11.1. A Solid State Logic mixing console, an industry standard in recording studios around the world. Its flexibility and computerized automation system makes it a powerful tool for mixing records from multi-track sources.

Gain Stages

When using any mixer, a primary objective is to get the signal through without adding unnecessary noise or distortion. To do this, one must pay careful attention to each of the "gain stages." A gain stage is wherever there is a chance for attenuation or amplification; simply put, wherever you have a volume pot or fader. "Unity gain" is a term that means a particular gain stage is set to pass the signal through without turning it up or down. Many faders indicate this setting with a darker line, or an indication of "0 dB," dB being the abbreviation for "decibel," which is a widely used unit of measurement for sound levels, devised by engineers of the Bell Telephone Laboratory.

Controls

While there are many brands and designs of DJ mixers with a vast array of features, there are four basic types of controls (Figures 11.2–11.5).

Fig. 11.3. The pot, a rotary control which adjusts audio signals by spinning clockwise or counter-clockwise.

Fig. 11.2. The fader which controls audio by sliding up and down or from side to side.

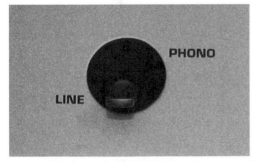

Fig. 11.5. The button, which turns audio on or off by being pushed. Some buttons have an up or down (in or out) position, others light up or change color when engaged.

Fig. 11.4. The switch, which turns audio on or off by clicking back and forth or up and down.

The following are the specific controls that are common to most DJ mixers, and what they do (Figure 11.6):

Gain/Trim: Controls the input level at the line in or phono pre-amp stage, before the signal is sent to the channel fader. This pot should be down all the way when connecting

Fig. 11.6. The DJs eye view of the Numark PPD-01 DJ mixer.

cables to the corresponding input. Ideally, the gain control should be adjusted so the channel's average signal meters at 0 dB when the channel fader is set to zero (unity gain).

Channel fader: Controls the volume of the signal coming through the individual channels. Scratch or "performance" mixers often have an adjustable slope control and a reverse or hamster switch for the channel faders (for more on hamster switches, see Chapter 19).

Crossfader: Fades between and blends the signal from two separate channels before being sent to the master output. Will usually have an adjustable slope control and reverse switch.

Crossfader tip

For the turntablist, the "feel" of the crossfader is one of the most important and personal aspects of a mixer. There are a variety of crossfader types: mechanical, magnetic, Voltage Controlled Amplifier (VCA), optical, and others. As with choosing any musical instrument, find a mixer that feels right for your style of playing.

Some crossfader wear quicker than others, eventually "bleeding" the signal through from both channels, even when you have the crossfader all the way over to one side. If you are performing regularly, carry a replacement crossfader and know how to switch them out yourself (this information is usually spelled out in the mixer's manual).

Channel switch: Usually switches between phono and line inputs. Often used by turntablists as an "on/off" switch, giving a more sudden attack than the channel fader or crossfader.

Channel EQ pots: Amplifies or cuts specific frequency ranges. Usually divided into: treble (or high) frequencies, midrange frequencies, and bass (or low) frequencies. This is known as "equalization," abbreviated EQ.

EQ tip

While recording consoles' EQ controls typically let you cut or boost between 12 and 15 dB, professional DJ mixers often let you cut specific frequency ranges entirely out of the mix.

This lets you cancel out the entire low end of a record (the bass, and most of the kick drum), or cancel out everything *but* the kick and bass; both are techniques that can be used effectively when segueing between tracks.

When scratching, try cranking the treble all the way up and killing the bass: this cuts record handling noise while giving the sample the power to cut through the phattest of beats.

EQ kill switches: Drastic EQ attenuation instantaneously available in switch form (see EQ tip).

Master volume: Adjusts the level going to the main outputs. Ideally, if the mixer's gain stages from the gain/trim control through the channel fader are set correctly, the master volume control can be set at zero (unity gain), for optimal fidelity.

Cueing/monitor controls: Adjusts the level and mix of what is being sent to the headphones (for a more detailed discussion, see Chapter 15).

Booth volume pot: Adjusts the level going to the booth outputs, which may be connected to an amplifier and monitor speakers located in the DJ booth (see Chapter 15).

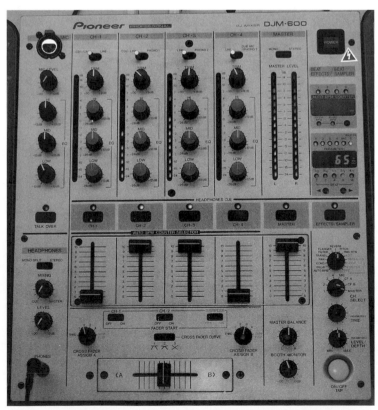

Fig. 11.7. The Pioneer DJM-600 includes an extensive array of features, including an onboard effects section, and a digital sampler.

Crossfader contour pot: Adjusts the slope of the crossfader as it fades between two channels. A gradual slope is generally appropriate for mixing, while a steep slope is usually preferred for scratching, especially for techniques like the crab.

Crossfader assignment switches: On mixers with more than two channels, it's necessary to assign the channels you want the crossfader to effect.

Headphone jack: Sometimes located on the front of the mixer. Usually a stereo (TRS) 1/4-inch jack; a few manufacturers have started to include an 1/8-inch stereo jack, due to the proliferation of headphones with this connection.

Microphone input: Often a 1/4-inch phone plug on DJ mixers, and XLR cannon plug most everywhere else. The mic input is often located on the front of the mixer, sometimes on the top, occasionally on the back.

Crossfader reverse switch: Reverses the direction of the crossfader so when the crossfader is over to the left, the channel on the right is heard, and vice versa. Sometimes labeled "Hamster" (see Chapter 19).

Inputs and Outputs

Phono inputs: Leads to a phono pre-amp, which includes the RIAA EQ curve, plug your turntable in here. The connection type is two Radio Corporation of America (RCA) jacks, left and right (stereo), red represents the right channel, white represents the left channel.

Ground: This post takes the ground wire from the turntable. If your turntable requires grounding via a ground post it will have an extra wire; connect it to this post by slipping it against the mixer's chassis and tightening the thumbscrew. Most modern electrical devices (including modern turntables like the TTX) have a ground connection built in; older turntables do not.

Line in: These high impedance, line-level stereo RCA inputs are usually used for hooking up CD players, Mini Disk players, and laptop interfaces. With the right cables (male 1/4-inch to male RCA), these inputs can also accommodate synthesizers and drum machines.

Aux input: High impedance, line-level inputs, usually RCA but sometimes 1/4-inch. Basically another line input, meant to accommodate "auxiliary" equipment like synths, drum machines, the line-level input from another mixer, etc. Of course, this can also be used to

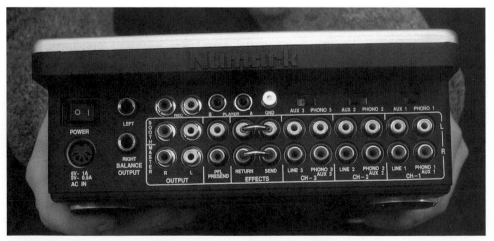

Fig. 11.8. The back panel inputs and outputs on the Numark Matrix 3 mixer.

input another CD, laptop interface, or MD player. On this mixer, any of the phono inputs can become an aux input by flipping the switch above the channel, bypassing the phono pre-amp.

Balanced output: This is your best quality stereo output, low impedance on either a 1/4-inch TRS balanced, or an XLR cannon. In most professional situations, this is the preferred way to connect to the next stage, be it graphic EQ, amplifier, or house system. Low impedance balanced signals reject interference and noise better than high-impedance signals, and are better suited to long cable runs.

The signal level of this output is affected by the master volume control.

Master output: Same program material as the balanced output, signal level also affected by the master volume control. The difference is this output is unbalanced, high impedance on RCA jacks.

Booth output: Signal level is controlled by the booth volume pot, can be hooked up to a monitor system in the DJ booth or other gear, high-impedance RCA jacks.

Record output: Signal level is not affected by the master or booth volume pots, but is affected by all other controls. Can be used for recording to CD, MD, or any format that accepts line level RCA inputs.

PFL presend: PFL stands for Pre Fader Level, and is a feature on some mixers which allows you to send a signal directly from an individual channel before it is affected by the channel fader or EQ. A typical application would be to send this signal to an external effects processor, then back into the mixer through the effect return (see effects loop, below).

Effects Loop

An effects loop is the most common way to incorporate outboard effects into the signal path of a DJ mixer (Figure 11.9).

Effect send: This is an output; you are sending the signal you want to process from the mixer to the input of the effects unit.

Effect returns: This is an input; the effected signal comes back into the mixer here from the output of the effects device.

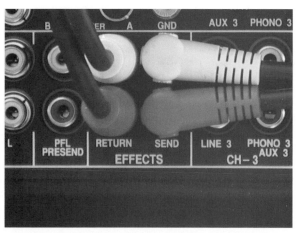

Fig. 11.9. An effects loop consists of a send and a return.

Digital Mixers

Digital DJ mixers function in much the same way as their analog counterparts, but the mixing and EQing is taking place in the digital domain. Upon entering the mixer, the audio is run through an A/D (analog to digital) converter, where the waveform is sampled and converted into Pulse Code Modulation (PCM) digital data, essentially a series of 1s and 0s like those found on an audio CD. One potential advantage of digital mixers is lower overall noise, especially in the EQ section.

A lot of engineers and DJs have strong feelings about their preference for either analog or digital signal paths. Use your own ears and listen critically to the sound of any mixer you are considering; there are excellent choices in both the analog and digital domain.

Effects

First off, effects, as cool as they are, are not the point. They can easily be overused. No amount of gratuitous effects usage can make up for a bad mix or a lousy set. Having said that, effects can be the icing on the cake. They can take things to the next level; make a good set into a killer set.

Onboard versus Outboard

The "board" here refers to the mixing board; outboard effects are external devices usually hooked up through the effects loop, onboard effects are contained within the mixer itself (Figure 11.10).

The obvious advantage of onboard effects is the "no muss, no fuss" aspect; nothing to hook up, plug in, or otherwise mess with. They also integrate with the mixer's controls, sometimes utilizing the crossfader and crossfader contour pot to adjust effect parameters. A digital mixer's built-in BPM detection technology can also make it easy set a delay, echo, or stutter-style effects to the tempo of the program material.

Outboard effects give you more choices; there are literally hundreds, if not thousands, of hardware effects processors to choose from. For DJ applications, effects that are easy to control in "real time" are preferable.

The Air FX lets you adjust multiple parameters of various effects by tracking the movement of your hand, sensing left and right (X-axis), front to back (Y-axis), and overall distance (Z-axis). For DJ use, the filters are especially effective, and the interface is intuitive and visually striking during live performance (Figure 11.11).

Fig. 11.10. The effects section of the Numark DXM06 digital mixer. Digital mixers often incorporate onboard effects into their design.

Fig. 11.11. The Alesis Air FX sends a beam of infrared light out of the top, then uses sensors in the dome which see the light reflected off of your hand.

The Kaoss Pad has inputs for a microphone or a turntable, and can also be used in conjunction with an effects loop. In addition to its filters and vocoder effects, the sampling functions can be useful for DJ applications (Figure 11.12).

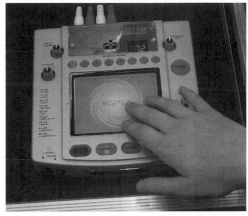

Hybrid Mixers

A recent trend is the analog mixer that also has a built-in computer connection, usually in the form of a USB or Firewire port. These allow digital audio to flow directly into a computer, where it can interface with any number of audio software programs, without the addition of a separate interface (Figure 11.13).

Fig. 11.12. The Korg Kaoss Pad senses the touch of your fingers on an *X/Y*-axis, and adjusts effect parameters accordingly.

As mixers continue to evolve, computers become more capable of mixing internally, making it likely that control surfaces which don't actually pass audio will continue to gain popularity (Figure 11.14).

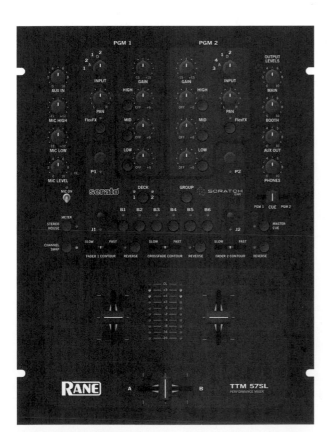

Fig. 11.13. The Rane TTM 57SL can convert the input from Serato Scratch Live time code control records into digital signals to control the SSL software without the need for additional hardware.

Fig. 11.14. Control surfaces like the one above provide dedicated controls for many software parameters, but do no internal processing of audio. Some have likened them to a very fancy mouse.

12

Video DJ Tools

Music and visual images have been artistically fused together for centuries through ballet, opera, musical theater, performance art, feature films, and MTV. While video Karaoke has been a popular pastime for years, many DJs are delving into more artistic live video endeavors.

As with audio, digital technology is enabling a significant leap forward for those interested in mixing video. Options include DVD turntables, laptop video mixing, and even hybrid systems that turn an analog turntables into video scratching and beat juggling machines.

Fig. 12.1. Scratch VJ ScientifiKent stakes out the cutting edge by controlling both music and video via Musical Instrument Digital Interface (MIDI) turntables using proprietary EJ vinyl, optical cartridges, and ScratchTV software.

The Visual Realm

Advances in digital video, video projectors, and flat panel video monitors have all conspired to make it possible for nightclubs and mobile DJs to delve into adding video to the mix. This can turn a venue into an even more transformative environment.

There are two distinct ways to mix video: by mixing music videos created for (and synchronized with) the music by the recording artist, or by creating new visuals to mash-up and synchronize with the audio content. Over an evening's performance, a VJ Video Jockey (VJ) may even incorporate both of these approaches.

The reason it is important to delineate between these two approaches is that with the first approach the video content will be coming off of the same device as the audio, with the second approach the video content will likely be created by a separate device. Some VJs work alongside DJs in a collaborative manner to visually interpret the music being mixed, others will also program the music as well.

DVD Decks

One of the first innovations to ignite the imagination of DJs in the visual realm was the introduction of the DVD turntable by Pioneer. Building off of their work in bringing to market the popular CDJ series of CD turntables, the Pioneer DVJ series caused quite a stir by putting video as well as audio under vinyl-like scratching control. In addition to vinyl emulation, the DVJs offer looping, digital pitch, and tempo control of up to $+70$ and -100 percent, and an interface familiar to anyone who has used one of the ubiquitous CDJ units (Figure 12.2).

In addition to video, the DVJ-1000 takes advantage of DVD's other features, including support of hi-resolution 96 kHz/24-bit audio, and the ability to store up to 1000 MP3 tracks on a single DVD-ROM.

The DVD format's dropping price point has made it more accessible as time goes on. The Numark VJ01 is a less expensive tabletop DVD player that lacks the DVJ's vinyl emulation wheel, but includes a built-in 5.6-inch tilt screen thin film transistor liquid crystal display (TFT LCD) (Figure 12.3).

Fig. 12.2. The DVJ-X1 was the world's first DVD turntable.

Fig. 12.3. The VJ01 includes pitch–control capability, and effects such as cue, loop, zoom, angle, and slow motion.

Numark also makes a rack mountable dual DVD player (the DVD01), with +/−50 percent pitch control capability and similar onboard effects and as the VJ01, without the built-in LCD.

Video Mixers

Adding video means that you're going to need a way to mix multiple video sources, and video mixers have traditionally been much more expensive than mixers in the audio realm. Manufacturers have been bringing to market less expensive video mixers specifically for VJs. Most include dynamic visual effects, which can often be placed under BPM control. As with music mixes, transitions are important in video mixing, so many mixers include a variety of variations for making transitions.

The Edirol V-4 mixer is an affordable four input/three output video mixer with more than 90 types of effects such as Still, Strobe, Multi Screen, Picture in Picture, and composition effects such as Chroma Key (blue screen) and Luma Key (black screen) capability for superimposing images. You can specify BPM via a TAP button, or a system called V-LINK, which interfaces with Roland musical instruments (Edirol is owned by Roland) (Figure 12.4).

When mixing with the Edirol V-4 and other similar video mixers, typically audio will need to be routed through an audio mixer, resulting in a two-mixer set-up.

The other option is to mix video and audio through a hybrid audio/video mixer, such as the AVM mixers from Numark.

These mixers feature inputs for four video sources, seven stereo audio sources and two microphones. The video capabilities include fades, 96 wipe patterns, picture in picture, two video effects sections, and chrominance and luminance keying (Figure 12.5).

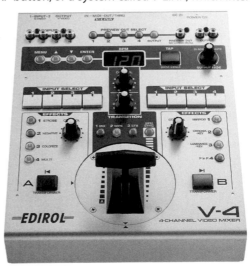

Fig. 12.4. The Edirol V-4 is loaded with 240 patterns for "wipes," to transition between sources.

Fig. 12.5. The AVM02 includes two separate DJ-style crossfaders for audio and video, which can be used independently or placed in linked mode operation. The joystick controls video fades and wipes.

Korg, whose KAOSS PAD is a popular effects box among DJs, jumped into the video fray with the "KAOSS PAD entrancer," which puts both video and audio effects under the fingertips of the VJ in a very kinetic and intuitive fashion. A unique feature of the KAOSS PAD entrancer is the way it simultaneously affects the audio and video signals, making for some magical synchronized output.

Korg also makes a simple four-channel video mixer named the "krossfour," and a self-contained piece of video hardware dubbed the "kaptivator Dynamic Video Station." The kaptivator is meant to be used as a video composition tool, and allows the user to sample, store, mix, process, and play back hundreds of video clips—plus live video—in real time, without the need for any additional equipment (Figure 12.6).

Fig. 12.6. Korg's kaptivator Dynamic Video Station features a familiar, drum machine-like interface, and can sample and play back up to 800 video clips using 16 pads. Real-time effects are controlled via rotary knobs, a slider, and a ribbon controller.

Video Software Options

As we've seen with the audio side of DJing, computers are making much of the hardware seem quaint by comparison. Unlike analog turntables playing vinyl records, VJing doesn't have years of tradition for practitioners to wax nostalgic about, meaning that laptops and computers may have an easier time taking over in a shorter amount of time.

Many software programs that let you create video mixes are currently available, and the number will certainly continue to grow. Many will allow you to download a sample version of the program to try it out before you buy. Some have hardware interface/controller options, and the ability to accept an audio input in order to pulsate effects in time with the BPM of whatever music you're playing.

A good place to start is with a program called NuVJ by the interactive visual technology company ArKaos in Belgium. NuVJ includes a two-channel video mixer that loops video clips in a variety of ways, with well-integrated effects and an intuitive interface (Figure 12.7).

Numark and ArKaos have teamed up to develop a NuVJ hardware controller that gives you the ability to scratch video, and spreads almost all of the software's features out onto dedicated pads and faders. NuVJ easily incorporates video clips from your computer's hard drive, and makes it possible to use a live video feed.

The results you get from NuVJ are pretty "artsy," and the program is easy to learn and to use, with a free demo version available for download at www.nuvj.arkaos.net. NuVJ is both Mac and Windows compatible, as is Arkaos VJ, another program from the same company, which excels at video creation for VJs.

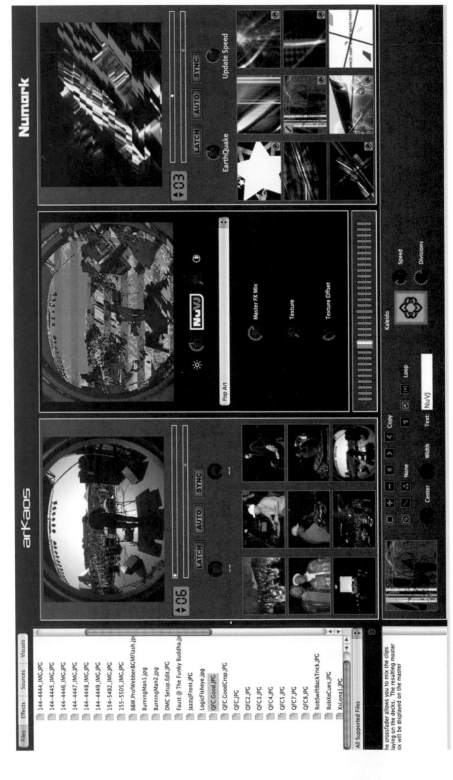

Fig. 12.7. The on-screen interface of NuVJ resembles a DJ mixer with a crossfader.

Another option for Mac OS X is Modul8 by a Swiss company called GarageCube. Modul8 lets you combine up to 10 loops at once, and effect them with a variety of effects, which makes the interface generally more complex, but not overwhelming.

Yet another current software option is "motion dive .tokyo," by Digitalstage. Motion drive has been in use in Japan since the late 1990s, and has formed a partnership with Edirol/Roland to develop and distribute a hardware console, the MD-P1, which will also interface with Roland's V-Link protocol. The software offers a two-channel video mixer with a crossfader, switches, and effects.

CUE, the live mixing software application, is capable of playing music video files in addition to music files.

DJ to VJ

The choice to add video into the mix is one that more and more DJs are making, for a variety of reasons. It's fun. It's creative in a similar way to mixing music. Mobile DJs report that they can charge more for a video dance, and in this competitive field, it can add value to what you do.

As digital media merges more disciplines into one, it's no wonder that DJs are once again at the forefront, exploring new tools and possibilities.

13

The Bizarre: MIT's DJ-I-Robot

If you came away from the movie *Good Will Hunting* with the impression that the Massachusetts Institute of Technology (MIT) is a pretty heavy place, then you can imagine what the brainiacs that populate MIT's famous Media Laboratory are cooking up.

Established in 1985, the Media Lab's unique synergistic amalgamation of scientists, uber-nerds, and artists has already given the world modern staples such as digital video and multimedia.

The Music, Mind, and Machine Group at the MIT Media Laboratory envisages a new future of audio technologies and interactive applications with the modest objective of changing the way that music is conceived, created, transmitted, and experienced.

Enter into this environment Chris Csikszentmihalyi (pronounced "chick-sent-me-hi"), Assistant Professor of Media Arts and Sciences at the Media Lab, who has told BBC news:

"We're trying to make human DJs obsolete, as far as possible. They're expensive, they're unreliable. If we can make this machine work, we'll give club owners an easy time".

Professor Csikszentmihalyi's Frankenstein creation, DJ-I-Robot, can scratch, beat juggle, backspin, and battle just like a human DJ, but it can also perform super-human feats, such as simultaneously spinning three platters at speeds of up to 800 rpm.

Csikszentmihalyi and his team are currently increasing the number of DJ-I-Robot's platters to eight, which may help it crush world-champion DJs like the steam engine killed John Henry, or like the computer named Big Blue opened up a hard drive of AI whup-ass on human chess grandmaster Garry Kasparov.

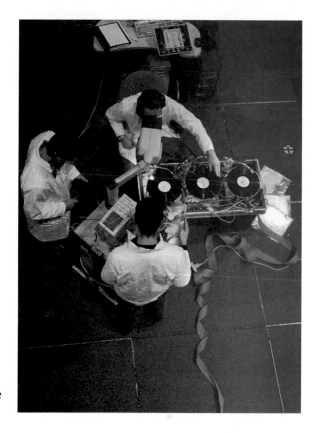

Fig. 13.1. Professor Csikszentmihalyi and his crew in the lab with DJ-1-Robot.

At the lab, they refer to the robot as the "DJ killer app." A program designed to record, store, and recreate the movements of DJs who scratch into its sensors is known as the "Soul Stealer."

But before you sell your decks in despair, you should know that Professor Csikszentmihalyi is not some Transylvanian mad scientist on an elitist ego trip bent on ridding the world of scratch DJs, despite what he says in the press. Chris is actually a slightly overeducated Hip-hop producer and fan from Chicago, who is skillfully playing the role of devil's advocate in order to explore with refreshing relevance issues of man versus machine, and technology colliding with art.

We talked in his cluttered office at the MIT Media Lab, which has windows overlooking a huge space, quietly bustling with graduate students, some of who were writing code for DJ-I-Robot's new interface.

Try, if you would, to describe DJ-I-Robot as you would at a dinner party, to someone that doesn't have a background in this sort of thing.

I haven't been invited to dinner parties in a while, because I keep trying to explain it! Basically, right now, it has three turntables, which are held together on an information network, called an RS 485 network, which is used a lot in industrial machine protocols, factory automation, that kind of stuff. That network is also connected to a PC running proprietary C++ that we wrote to control the turntables.

High-speed motors, accurate to 1/6400 of a revolution, drive the turntables. They have encoders that run at about 48 volts, so they're able to spin from 0 to 800 rpm in a fraction of a second.

It has such remarkable acceleration and consistency that you can mark a high-hat at the beginning of a record, and when the stylus reaches the end of the record you can say, "Go back to the marked high-hat."

It will then spin at 300 or 400 rpm and find that exact high-hat.

One of the things that we first realized we would have to imitate was the backspin, which I would say is probably the most important thing in DJing—even more so than scratching.

Our tests indicate we can probably backspin uninterrupted on our system for about four and a half months. If you do a backspin overnight, when you come back, there will be white all over the record from the plastic that's been eroded.

The pressure on the needle is pretty good so that it won't skip?

Yeah, it's pretty good, and it's doing 800 rpm, which is much faster than most DJs can turn. When it's really chewing into the vinyl, it's pretty serious.

As we went on with the system, we developed a way to compose something that DJs would call beat juggling. Essentially, it's the ability to remix two or three different records and to pre-compose it, by saying:

These are different parts of the song. I want you to go play these four measures on this turntable. Then, after that, play these two measures on this turntable. Then, after that, play half a measure on this turntable. Then, half a measure on that turntable, and back and forth.

What do you say to people who question why you don't just manipulate digital audio inside of a computer to achieve the same editing?

It was important for us to have an analog mentality.

I think turntablism uses turntables and records for a lot more reasons than faith in analog.

I think that a lot of DJs couldn't care less about the quality of sound. It's the tactility and the physicality of the record that are really important.

Nonetheless, it seemed that if we were going to be doing this, we should maintain a fully analog signal path. I had to hunt around for a long time to find this particular generation of technology where there were MIDI-controlled analog mixers. I found a couple of those on eBay, and that's what does all the mixing in the system. It's Mark of the Unicorn MIDI Mix 7, or something like that (Figure 13.2).

We premiered the system as it stood in 2001.

Fig. 13.2. The DJ-I-Robot's analog signal path.

The problem was that we could compose an hour-long, beat-juggled performance with no problem, but the chance of it skipping after 10 minutes was 100 percent, and once it skips, we're screwed.

It would be easy for us to put incredibly expensive sensors on to figure out if it had skipped, but it's always been important to keep it kind of "street."

The skipping was enough of a problem that we couldn't do a beat juggling set for more than five minutes, without really worrying.

So, some of our early concerts were exercises in nausea, watching this system that's completely autonomous going out on its own, and just being terrified that it's about to skip.

Kind of like a parent watching a child . . .

Exactly, like watching their kid's piano recital.

So we said, "Okay, number 1, the composition process is very different than what DJs do." We began to think of the system more as a computer-assisted DJ and started developing, for example, a performance interface. This interface allows you to do many things—for instance, scratch by remote control.

You can also scratch the performance interface and make all three turntables scratch in perfect synchronization. Or you can record the motion of the scratch and play that motion back over different parts of the record.

Have you interfaced with prominent DJs to come in and do some scratches?

Our goal is to create a database of scratches. We have a piece of software that's specifically made to record someone's gestures and save it into this database. We partly want to immortalize DJs' techniques, so later, people will be able to scratch their signature scratch over new music that's just come out.

Another thing you could do is take two scratches, one by DXT and another by QBert, and then morph the two to see what their children would scratch like.

That piece of software is actually called the "Soul Stealer."
[Laughs]

Do you tell the DJs about that?

It depends on if they like me already or not. Next June, we're going to be doing a tournament in LA, with Kool Herc, Theodore, and a bunch of really famous guys. That's where I'm going to start rigorous documenting of these people's scratches. I kind of want to do it with the old school first.

I understand you've had some interaction with DJ Spooky?

For Spooky, we originally did a remix for the title track from his *Optometry* LP. That should be coming out any time. He had, I think, four or five different DJs remix that song, and the robot was one of them.

Tell me about the process you went through to remix "Optometry."

We spent a lot of time just setting up different routines and different moves. We did some modeling of what it was like when we had good people scratching into the system, and then we'd reproduce those moves with slight variations. The last stage was traditional remixing, where we brought those elements in as multiple tracks and also added a lot of stuff.

When you have somebody scratching into the system, are you able to capture their crossfader technique?

Yeah, we basically have a mixing board that has three crossfaders. But the crossfaders aren't actually controlling the volume; they're going to an analog-to-digital converter. So we can have the crossfader controlling volume, but we can also have it controlling other things.

Like, if we have one knob scratching all three turntables, we can set their volume faders up and down. This essentially reproduces an echo, varying the amount of delay from when you do the scratch to when the turntable actually does it. So, the volume faders can become mappable to any parameter (Figure 13.3).

Tell me a bit about your background.

I was a college dropout. I started a job in anthropology, and dropped out after a couple of years from Reed College. I went back to finish my bachelor's degree at the Art Institute of Chicago, mostly interested in sculpture and art, and technology. I also did some film.

Fig. 13.3. The dedicated interface for the current DJ-I-Robot includes three crossfaders, and a knob to control the movements of any or all of the platters.

I got into music video work for about a year and a half, after I got out of college. I was also producing some Hip-hop with a band called "He Walks Three Ways," in Chicago. I was their token white boy. So, I had a studio and was doing some electronic music.

What kinds of things were in your studio?

Samplers, a Mac running a couple of different sequencing programs, and a couple of turntables. We actually did some pretty good Hip-hop.

One of the things we were talking about, one day, was this idea of this gigantic junk heap with all types of things that we were interested in, that we could stop-motion animate and have move around. One of the things that we'd have this junk heap do is grab a turntable and scratch, as it would walk across the room. We were really interested in the idea of scratching being automated by this amalgamation of our interests.

Eventually, I went to grad school at UC San Diego, where there was a kind of obscure artist named Harold Cohen, who developed a robotic painting machine over a period of 30 years. There are a couple of books published on the AI behind it and his theories about painting. I disagreed with almost everything he said, because he really believed that his system was painting.

I believe that I have yet to see a music technology that can, in some way, approximate taste, human values, or judgment in music.

The grad program was primarily people involved with more traditional media, but there were a few of us that were doing active work with computers. There are some things about computers that I feel are radically different than any other existing form of media—the main difference being that they can sense the world, make some kind of decision, and affect the world in their own ways.

There came a point when I realized that I wasn't clubbing enough anymore; I was spending all my time hacking and doing electronics.

I thought, "How can I still do the work that I want to do, but re-engage with this community that I haven't been engaged with in the last few years?"

At that point, I thought, "Maybe, what I'll try to do is build a DJ system that would somehow be computer controlled." I built a very quick prototype in late 1998, early 1999. The prototype was terrible, but it was good enough that you could already hear sounds that hadn't been heard before. Strange little whines and groans from the turntable that no one had ever made.

I was telling a colleague, "I'm going to build a robot DJ next," and a jazz bassist named Curtis Bond looked at me and said, "You're going to be famous!"

How has the DJ community responded to this project?

Some high-profile DJs, like Spooky, have been really nice to us. It's funny, because there are a lot of DJs who are not at all interested in the idea. They probably see the goal as being to replace DJs, and it absolutely isn't. On the other hand, some people have really been into it.

To me, one of the defining moments was really early on in the process. We were just getting the physical system working. It basically runs from motors that have what are called rotary optical encoders in them, so if you put voltage in the motor, it will spin. Then the encoder will say where it's going.

So, you put a little computer in between those two things, and by looking at the encoder, it creates what is called a feedback loop, similar to the way our nervous system works. If the feedback loop isn't strong enough, then the motor won't get quite where it needs to go. If it's too strong, it will overshoot, and it will bounce forward, and bounce back, and bounce forward, and bounce back.

It's called a "sprung math." Early on, we had the math a little bit wrong, so we would tell the motor to go one place, and it would go a little too far. Then it would bounce kind of like a spring.

We had this DJ come in to give us some feedback on our system. We were explaining that this was really a mistake; the motor shouldn't be overshooting and making this funny sound.

He looked at us and said, "That's not a mistake! That sounds great! I've never heard that before!" He came back two weeks later and was really excited because he had managed to figure out how to make that sound! That, to me, was a great moment.

Why set out to replace the DJ? What response are you hoping to elicit?

Looking at Deep Blue versus Kasparov and these other kind famous cases of human/machine conflict, I'm always fascinated by the way it comes into the public imagination.

To me, the formulation that was happening in DJing was analog versus digital.

Turntables are selling more and more, and records aren't disappearing the way that they were supposed to in the 1980s, when CDs came out.

The music industry still feels that digital is inevitable, and who knows, they might be right. Part of the reason it may be inevitable is because the industry is putting so much energy into making it inevitable. As it is, all this research is going into developing new DJ systems that are CD or MP3 based.

What really interested me is that most of them ignore the reason why records are so popular.

The records themselves have memory, on a certain level; they have a nostalgia value and kind of a fetish value.

You can hunt down incredible records that no one else has access to, if you're going to a great neighborhood with old stores. But also, they're physical. You can look at them and see information, check out what's going on with them.

Interface designers for digital systems haven't really addressed these issues.

They have really cheesy interfaces without any kind of action feedback. You can't see what's coming up ahead, except on a sound-wave representation, which isn't a physical-time based representation in the same way. Why are people trying so hard to replace records, rather than thinking about a way to produce records very cheaply, or produce records in the same way that people burn CDs at home?

You been quoted in the press saying, "DJs are lazy, they sleep late, and are addicted to cocaine. They spread sexually transmitted diseases." Why do you make statements like this, if you're not really trying to replace DJs?

I'm stating the process the way that many people who produce new technologies do. These new technologies are not trying to add something. They do it in a knee-jerk reaction by trying to destroy something and put something over it. When digital came out, people had to pretend there were all these huge problems with analog.

But the tactility of records is something that was never listed as an advantage. It was only the fact that they scratched and warped, or whatever. The fact that you could look at them and manipulate them with your hands was never listed as an advantage. So, what happens is, technology people overplay their hand completely.

To me, the only way as a society that we can ever get it under control is to stop trying to denigrate existing technologies and to be a little more clear-sighted as to why we're building a next generation of machines.

So, when I do interviews in character, I'm trying to be a little more obvious than many of my colleagues, who don't feel the way that I do. Be a little bit more honest about

Fig. 13.4. "Rather than replace the vinyl, I figured I would try to replace the DJ"—Chris Csikszentmihalyi.

what I'm trying to do than they are, and thus, get people a little bit suspicious of that kind of terminology.

This radio interview I did in San Diego was great because someone called in and that was exactly his point, "Why are you MIT guys always trying to replace people with machines?"

Did you stay in character for the whole radio interview?

Oh, totally, totally. Actually, I talked so much that I managed to dominate almost the entire hour. It's very sad, how easy it is for the press to pick up those narratives, about technology. I mean, they're narratives that we've learned since birth. "Well, this little sucker never takes a coffee break." The term is just used over and over again.

Where does DJ-I-Robot go from here?

Right now, we have three turntables and a lot of development has been happening. Without much problem, we can expand the system up to about 128 turntables. We're hoping, by the June concert, or by a couple concerts in New York or one in LA, to have at least eight turntables going at that point.

The thing is, with three turntables, you can set up a backspin on two and scratch on the third, and so the computer-assisted side of it allows them to do things. We don't want to end up with eight turntables in a situation where everything has to be pre-composed, because eight is far too many for one person to really control. Although, Christian Marklay did 99, but there wasn't the level of direct interaction that there is with turntablism.

The trick is somehow to come up with a completely different system for running them, where large-scale changes can be made very quickly.

After eight, I don't think I'm going to go much further. If we don't get into the DMC with eight turntables, we're sure it'll never happen.

Fig. 13.5. MIT Media Lab graduate students work on writing code for the robot DJ.

PART III

The Skills

14

Approaching the Turntable As a Musical Instrument

Whether you mix in nightclubs, rock your mobile rig at dances and weddings, or scratch with a live band, your turntables (and mixer and other peripherals) are your instruments. They are your interface with the music, and you can maximize your effectiveness on them by applying what's in this chapter.

When approaching any musical instrument, one of the first issues is how best to use your body to most effectively interface with it. Like pole vaulting, the martial arts, or playing the cello, there are physical realities to deal with when approaching the turntables, including balance, gravity, and the most efficient use of your skeletal and muscular structure.

Physical Considerations

There's plenty of science involved in understanding how we are put together neurologically. Researchers are striving to better understand the electrical pathways that are formed to coordinate muscle and brain, intention and execution. This much is already indisputable: it is much easier to learn good physical habits than unlearn bad ones.

The following is an application of the body of knowledge that has developed around the pedagogy of musical instruments, as well as specific research about playing the turntable as a musical instrument. By applying both, we can employ practicing and performing techniques that will lead to maximum physical versatility, and long-term health benefits.

Table Height and Wrist Position

To develop efficient body position, first consider the height of your turntables and mixer. In order to facilitate a neutral position for your wrist in relation to the crossfader and the records, the platters of your turntables should be approximately even with your navel.

This is easily accomplished using an adjustable table (Figure 14.1).

If you do not have an adjustable table, you can use road cases to adjust the height of your turntables. Placing your turntables on top of your closed road cases gives you the most height (Figure 14.2).

Fig. 14.1. An adjustable DJ table, like this one from Grummund, lets you find a comfortable height.

Fig. 14.2. Road cases can raise turntables up when using a short table.

Fig. 14.3. Suspending the mixer has the added benefit of giving you space underneath for an effects unit such as the Korg Kaoss Pad.

For a few inches less height, keep the turntables in their road cases and use the other half of the cases as a base. This method allows you to suspend your mixer between the two road cases if your mixer has "rack ears" (Figure 14.3).

Hand Position

Dr. Majera T Majidi (AKA Doc Maji to his patients) is a chiropractor in Scottsdale, Arizona, who specializes in treating DJs, helping them utilize their bodies to perform at their peak while avoiding injury. He's studied the biomechanics of competitive DJs' shoulder, wrist, and finger movements for the past several years.

"There's a certain position of the wrist, we call it 'open chain neutral.' We find that DJs who use the crossfader in that position show the most speed and efficiency."

In order to put your wrist into the open chain neutral position:

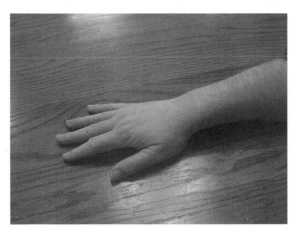

Fig. 14.4. Place your hand on a table, resting flat.

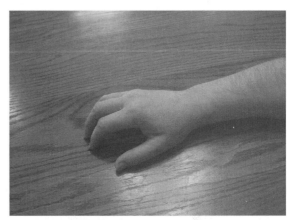

Fig. 14.5. Slightly curl your fingers.

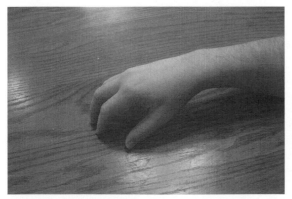

Fig. 14.6. Slightly raise your wrist.

Fig. 14.7. Move your fingers to the side, creating a 'V' in the area between your thumb and fingers.

Fig. 14.8. Keep your hand, wrist and fingers relaxed as you move your open chain neutral position to the mixer.

Fig. 14.9. Keep the curvature of your fingers and wrist in the same gradual curve as you move to the record hand as well.

"Open chain neutral position gives the largest radius in the carpel and ulna tunnels for the tendons that control the fingers to glide easier," explains Doc Maji. "This translates into more precision, more speed, and more accuracy."

Check out the following close-up photographs of QBert, Rob Swift, Logic and Perceus. Notice the relaxed, natural position of the wrist (Figure 14.10–14.13).

Fig. 14.10. DJ QBert.

Fig. 14.11. Rob Swift of the X-ecutioners.

Fig. 14.12. DJ Logic.

Fig. 14.13. US DMC Champion DJ Perceus.

Stretching

Many doctors who specialize in working with musicians recommend stretching before and after practicing, and Doc Maji recommends the same for DJs. Stretching can help you play longer without pain, and keep you from injuring yourself.

As with any stretching regimen, train—don't strain. Start slow and easy, and if something feels uncomfortable, don't do it. If you are physically challenged in any way, please consult with your physician before doing the stretching exercises below.

Doc Maji recommends that DJs should focus on stretching the pectorals, sternal chydomasteroid (SCN), trapezious, and the forearm muscles.

Stay relaxed as you do the following exercises. Start by taking a few deep, cleansing breaths.

Your pectorals are the large muscles in your upper chest.

In order to stretch your pectorals:

1. Stand in a doorway.

2. Extend your arms up above your head.

3. Grip the top of the doorway with your hands.

4. Carefully let most of your weight hang forward for five to fifteen seconds.

The SCN is the large muscle in the front of your neck.

In order to stretch your SCN:

1. Lie down on a bed on your back.

2. Carefully extend your head and neck off the bed.

3. Allow the top of your head to become parallel to the floor, stretching the SCN.

4. Turn your head to the right while you're stretching, and hold it for five to fifteen seconds.

5. Turn your head to the left while you're stretching, and hold it for five to fifteen seconds.

6. Carefully bring your neck and head back onto the bed, and wait a few seconds before slowly getting up.

The trapezius muscle is the muscle you use to shrug.

In order to stretch your trapezius muscle:

1. Turn your head to one side.

2. Bring the ear, which is facing forward, down toward your chest.

3. Hold for five to fifteen seconds.

4. Turn your head to the opposite side and repeat.

Your forearm muscle is located between your elbow and your hand.

In order to stretch your forearm muscles, start by extending your wrists backwards using the following exercise:

1. Extend your right arm forward, shoulder high.

2. Face your palm forward, as if signaling someone to halt.

3. Stretch your fingers backwards toward your face.

4. Hold for five to fifteen seconds.

5. Repeat using the left arm.

In order to stretch your forearm muscle in the opposite direction, extend your wrist forward using the following exercise:

1. Extend your right arm forward, shoulder high.

2. Face the back of your hand forward, as if showing off a new ring.

3. Stretch your fingers forwards toward your stomach.

4. Hold for five to fifteen seconds.

5. Repeat using the left arm.

Beating the Burn

As you practice, efficient body position is extremely important to reach your full physical potential, and to keep from injuring yourself.

"DJs should be aware that if they're doing a lot of crabs and flares and they start getting burning or tingling sensations in their hands, they're basically overdoing it," advises Doc Maji. "Despite what most people talk about, carpal tunnel syndrome, in DJs it's a different problem: *pronator terres* syndrome."

"When you're doing a repetitive flare or crab exercise, the muscle's going to swell a little bit. When that muscle swells, the nerve that goes through the muscle to supply sensation to the hand gets compressed, like stepping on a hose. When there's less fluid in the nerve, that's where the tingling comes from."

If you feel a burning or tingling sensation in your wrist or hands, stop for a minute and shake your hands out in a downward motion, letting circulation replenish the fluid, called acetylcholine, inside the nerve.

Digital Considerations

All of the above hold true whether you are using vinyl records, a digital turntable, or even a dual-well CD unit with a separate control surface. If you are mixing or scratching, you'll do better if you use your body in such a way that you avoid injury and maximize efficiency (Figure 14.14).

Lately, I've seen DJs who are incorporating laptop interfaces placing their laptops in impossibly uncomfortable positions. Craning your neck to peer into your computer's display may cause you just as much trouble as having your other

Fig. 14.14 Good height and body position is just as important when using digital gear.

Fig. 14.15. DJ RaeDawn using a laptop stand to elevate his laptop display to a more usable height.

gear placed poorly. There are a variety of laptop stands that may aid in finding a set-up that works for you. Keep in mind that if your set-up is a bit uncomfortable, by the end of a four-hour set you are going to be in pain (Figure 14.15).

Headphone Concerns

Like rock musicians, DJs face high sound pressure levels (or SPLs) as a part of our profession. At least as DJs, we've usually got our hands on the volume knob.

Keep in mind that what permanently damages hearing is dangerously high SPLs over significant periods of time. Also be aware when setting headphone levels that back spinning in the headphones will increase in volume the faster you go.

There is another consideration having to do with headphones that most DJs never think about …

"I see a lot of sinus problems," reports Doc Maji. "I see it a lot more in the mix DJs, and I think it is because they are spending a lot of time with the headphone covering their ear."

"When you have your ear covered like that, the oxygen level decreases to the inner ear. There are always bacteria in our ears, and when the oxygen level decreases, bacteria reproduction rates increase; the studies show 700 times more per hour. And because the Eustachian tube has a connection to the sinus as well, the bacterium has a travel pathway to the sinus."

In order to keep free from headphone-related sinus infections, intermittently take off your headphones and let your ear breath. Doc Maji also recommends cleaning your ears at night before you go to sleep.

15

Mixing Skills

Mixing is the bread and butter skill of the DJ. Done poorly, it can be excruciating for the audience. In the hands of a master, mixing has the potential to rise to the level of an art form.

"I saw Kid Capri rock a party in 1998 and it blew my mind," Rob Principe, CEO and President of Scratch DJ Academy states. "The way he manipulated the crowd was awe-inspiring. He took 1500 people and turned them upside down. It moved me … just like a good book or piece of art would move somebody. It was after that inspiration that I truly believed in the power and reach of the DJ."

When DJ Radar was just 16 years old, he witnessed DJ Emil, his mentor and fellow Bomb Shelter DJ, mixing one such magical set:

It was in Tucson Arizona; a huge rave of about 2000 people. They had a main room, but Emil said, "I'll just take the back room today." He was rockin' it; his room was just packed. Everybody was dancing, even all the DJs who were there watching.

All of a sudden, one of his turntables went out. All the DJs said, "Uh oh. Party's over."

But Emil kept it going! He got on the mic and said, "Hey, one of my turntables went out, but you know what? Let's just keep this party goin'!"

He got everyone to start clapping to that beat off of the house record he was playing. That's how he mixed; he would mix off of everyone's claps. He would switch records and keep time. He kept the party going for two hours! Everyone in the main room came over there to watch him mix on one turntable and the whole place was just blown away.

Experience and confidence are essential, and those are two

Fig. 15.1. Mixing live can involve a great deal of skill, including the skill of making it look easy.

things that you can't get from a book—you'll need to acquire them as you get out and mix in front of people. What we can do here is take a close look at specific skills, including basic (radio) cueing, slip cueing, beat matching, syncing records and CDs, musical considerations such as phrasing, and live remixing techniques. In Chapter 16, we'll apply these skills to building a set, reading the crowd, and developing your own style.

Cueing

Cueing is the obvious place to begin. If you're already adept at cueing, you may want to skip ahead. If you're starting out, you're in the right place.

For the DJ, cueing is the act of getting a piece of music ready to play, usually without the audience hearing it. This is also known as "cueing up" a record. You might be cueing up the beginning of a track, the last chorus or climax, the breakdown section, a drum groove, or the "Ahhh" sample to scratch with.

Regardless of what you're cueing, there are both visual and audible aspects to cueing. Let's check out the visual aspects first.

The visual aspects of cueing vinyl include the following (Figures 15.2–15.6):

Fig. 15.2. Counting the tracks, which are separated by flat segments.

Fig. 15.3 and Fig. 15.4. Marking the label so the track starts at 12 o'clock, or with the mark pointing at the stylus.

Fig. 15.5. Looking closely at (or reading) the grooves to figure out where the breakdown is.

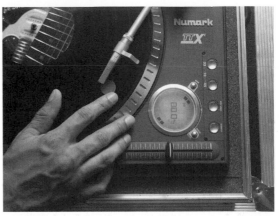

Fig. 15.6. Marking the exact location of a cue with sticker.

The visual aspects of cueing CDs include the following (Figures 15.7–15.10):

Fig. 15.7. A digital read-out of the track number, and the elapsed or remaining time of the track.

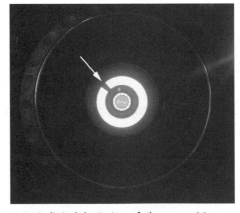

Fig. 15.8. A digital depiction of platter position.

Fig. 15.9. A low-res digital depiction of the track's waveform.

Fig. 15.10. On the Numark CDX, the vinyl record on the platter can be used to cue, just like with a turntable.

The visual aspects of cueing on a laptop may include the following (Figure 15.11):

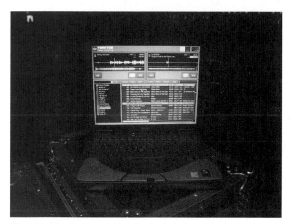

Fig. 15.11. A more detailed digital depiction of each track's waveforms, and a bin of available tracks.

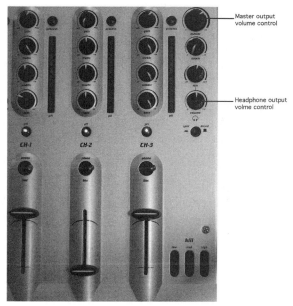

Fig. 15.12. DJ mixers have the capability to route source material to two signal paths: one leads to the main outputs, the other to the headphones.

## Headphone Cueing

Most of the time, cueing is done in the headphones while the previous record is playing for the audience. Therefore, you must determine how to listen to a source in the headphones without sending that source's signal to the main outputs of your mixer (Fig. 15.12). Practically all current DJ mixers include cueing controls, and while there are variations in how these controls are labeled and a few other specifics, their function is essentially the same (Figures 15.13–15.15).

Fig. 15.13, Fig. 15.14, and Fig. 15.15. Headphone volume controls on various DJ mixers.

When you're working with a new mixer, first locate the volume control for the headphones. This could be labeled "cue gain," or "phones level," or may have a headphone icon next to a volume pot. *Always* begin headphone monitoring at a low level to avoid damage to your ears or your headphones.

Big-time club DJs get to specify what mixer they want the club to have in the booth when they come to play. Before you get to do this, you'll probably need to be able to handle the cue systems on a variety of mixers.

Look to see how the cue system's controls are labeled on the mixer you are using. There are lots of variations, here are a few of them (Figures 15.16–15.18).

Fig. 15.16. "PFL" stands for "Pre Fader Level." This button lets you send a channel's pre-amplified signal to the cue section (thus hearing it in the headphones) even if the channel fader is all the way down.

Fig. 15.17. This "cue" rotary pot controls the proportion of channel 1 verses channel 2 signal sent to the headphones.

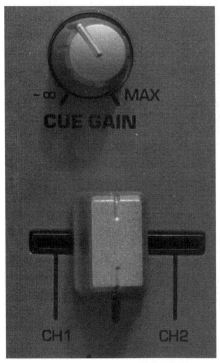

Fig. 15.18. Some two-channel mixers, like this Numark DMX 06, include a mini crossfader, which lets you select and blend between sources in the headphones independently of the master crossfader.

Splitting Your Brain

When cueing up records during a set, most DJs will slide one ear out of their headphones in order to keep an ear to the sound coming out of the main system, while also listening to the record being cued. This will help to keep you from accidentally sending the channel you are cueing to the main outputs (an unfortunate but common early mistake), but also requires you to listen to two things simultaneously. Some DJs describe this as "splitting your brain," and while it's a bit disorienting at first, with practice it will become second nature (Figure 15.19).

Fig. 15.20. The volume pot for the DJ booth's monitor system.

Fig. 15.19. Cueing a record with one ear out of the headphones lets you hear both the record that is playing for the audience, and the one you're cueing.

When you practice cueing, have the main outputs of your DJ mixer driving speakers in your practice space, and practice sliding in and out of your headphones and listening to two things at once. Challenge yourself by mentally tuning out one of the two records and listening to just the bass of the record coming from the speakers. Then have your brain tune out the record coming through the speakers and just listen to the snare of the record in your headphones.

If this seems impossible, consider the following scenario: you're at a party, and right in front of you is the world's most boring person telling you about something you have no interest in, like the eating habits of their pet cat. Behind you, a very racy conversation begins about your girlfriend (or boyfriend, depending on who you are). Believe me, you would have no problem tuning out the world's most boring person and catching every detail of the scandalous conversation going on behind your back. Try imagining that the piece of the record you are trying to isolate is this conversation.

Many large venues will have a monitor system in the DJ booth, which will also have a separate volume control on the mixer. This usually mirrors what is coming out of the main system, which may be difficult to hear accurately from the stage or the DJ booth (Figure 15.20).

It's important, especially when you're first starting to play out, to get some practice/ sound check time in on a big system. At high volume in a large venue, sound from the main speakers (especially low-frequency information like kick drum 'n' bass pumped through massive subwoofers) may arrive at the DJ position significantly later than it hits the dance floor. The monitor system in the booth (if there is one), the configuration of the venue's set-up, and the room's specific acoustics are all factors in this phenomenon. Practice, and a good, enclosed pair of headphones will help a great deal.

There are times when you'll want to slide both ears into the headphones, to isolate a record you're considering, or to more accurately hear the exact timing of two records you're beat matching. Some mixers have a "split" feature, which let you assign one track to the left ear and the other track to the right ear (Figure 15.21).

Fig. 15.21. The split button can help you keep track of what's happening on which record.

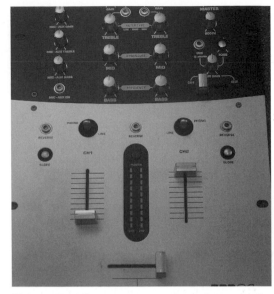

Fig. 15.22. Mixer set to cue channel one in the headphones.

Just realize that when you have both ears isolated, you have no audible way to tell what the audience is hearing through the main system. Therefore, you must be especially vigilant to visually check the mixer's controls to be sure you are sending exactly what you mean to be sending to the PA.

The Radio (or Basic) Cue

The radio or basic cue uses the Start/Stop (or Play/Pause) button to put the music in play, usually to play tunes one after the other. This method is used for live radio shows (thus the name), and other situations where beat matching and slip cueing are not called for.

Determine how to listen through the headphone (or cue) section of your mixer without sending the signal to your main outputs. Take the channel fader all the way down, or place the crossfader all the way over to the other side of the mixer, while assigning the channel you're cueing up to the headphones through the mixer's cue section (Figure 15.22).

Radio Cueing Vinyl

When handling vinyl, you want your hands to be clean, but not too dry. While you can't avoid it entirely, the less your hands come in direct contact with the grooves, the better. I'm quite a bit more careful with records that are hard to replace than I am with scratch records that are widely available.

First, we'll cue up side one, track one of any record in your collection. If your turntable has start-up and/or brake speed adjustments, be sure they're on the fastest settings available. Set up your mixer so that the channel you will be cueing is not coming out of the main outputs, but is being sent to your headphones. Put the record on the platter, let it spin and place the stylus at the beginning of the record. As soon as you hear the first sound through the headphones:

1. Press the Start/Stop button, stopping the record.
2. Place your middle finger between the label and the last track (Figure 15.23).

Fig. 15.23.

Fig. 15.24.

Fig. 15.25.

Fig. 15.26.

3. Rotate the record backwards, listening as you back up to the top of the track (Figures 15.24 and 15.25).

4. Notice the position of the label as it corresponds to the first sound on the track.

5. Move the record back and forth over the first sound a few times (thunk-a, thunk-a) to get comfortable manipulating the record.

6. Leave an inch or so of space between the stylus and the first sound (Figure 15.26).

The record is now cued up. On any decent DJ turntable, the platter should have stopped quickly when you pressed the Start/Stop button, so you shouldn't have to back up very far.

To check your cueing job, turn up the corresponding channel of your mixer, place the crossfader in the center, and press the Start/Stop button. If the record is not all the way up to speed when the first sound is heard through the speakers, you need to leave more space between the stylus and the first sound. Practice cueing in this way until you've gotten a feel for exactly how much space to leave with your particular turntable, and how much time passes between hitting start and hearing the first sound.

The next step is to practice cueing up one record in the headphones while another record is already playing through the speakers. Cue up a record on the left turntable using the instructions above and let it play.

With the crossfader all the way over to the left, place a second record on the right turntable and adjust the cue controls until you hear only the right turntable in your headphones. Slide one ear out of the headphones so you hear both records.

Cue up track two of the second record. Practice accurately placing the needle in the space between tracks one and two. It's easier to line this up visually in standard mode (with the tone arms to the right) than it is in battle mode (with the tone arms across the top), which is why most mix DJs set up in standard mode.

Repeat the steps above to cue the second record. Move the crossfader to the center, then play the second record at a time you deem appropriate (as the previous record begins to fade, for instance). We'll talk more about transitions in a moment.

Radio Cueing CDs

When handling CDs, you want your hands to be clean. Avoid getting fingerprints on the bottom of the CD, or it may skip and do other erratic things at inopportune times. Hold CDs by the edge, or stick your index finger through the center hole.

Track Three

Let's start by cueing up the top of track three of the CD that came with this book. Load the CD into one of your decks and adjust your mixer so the channel you're cueing is not coming out of the main outputs, but is being sent to your headphones.

Specifics will vary, depending on your CD player. Let's explore cueing up a CDJ-1000 in CDJ mode, where the deck acts more like a traditional CD player.

In this mode, the *Auto Cue* function sets your cue point for you by finding the first sound of a track and marking where the sound is. Most DJ CD players have a version of Auto Cue. This function is especially useful because it saves time cueing up a track.

To turn on Auto Cue and cue up track three:

1. Press and hold the *time mode/auto cue* button on the top left of the CDJ-1000 until *A.CUE* lights up on the *display*.
2. Press the *track search* button repeatedly until you get to track three.

The cue point for the track is automatically set, and the CDJ is placed in pause.

The only drawback to auto cueing like this is that you haven't actually checked the track before the audience hears it. This means you haven't made sure the gain is adjusted on your mixer, checked to see if the track has some silence before the music starts, or double-checked that the CD is indeed the right one. This could be okay if you are very familiar with the CD and track you are playing. If not, you may want to check.

To check the track:

1. Press Play. You should hear the track in your headphones.
2. Press the Cue button. The CDJ will locate back to the set cue point and playback will stop.

Turn up the corresponding channel of your mixer, place the crossfader in the center, and press the *Play/Pause* button. Playback of the track will start from the set cue point.

Basic Transitions

There are important factors that determine how well two records will work together, and how best to go about making a skillful transition.

The key and tempo of each song, along with the exact nature of the intro and outro must be taken into account. Crossfading between two songs can be immensely effective or disastrous depending on these factors.

As you gain experience, you'll find records that compliment each other beautifully, and lend themselves to overlapping in different ways.

If you're playing a mobile wedding gig, or some other event where you've been handed a group of requests and an order to play them in, you'll want to check out each record in the headphones while the previous record is playing. If the keys are consonant with each other, or there is a drum intro, you may want to crossfade between the records, or even beat match them (more on this later). If not, it may be most esthetically pleasing to start the new record immediately after the previous record has finished.

In order to get your feet wet with basic transitions, use the cueing techniques described above, and practice spinning some sets. Notice which transitions work and which ones don't, and get comfortable cueing, running the mixer, and splitting your brain. Once you get this down, it's time to move on to slip cueing.

Slip Cueing

When slip cueing, you're setting the next record into motion by hand, rather than using the Start/Stop button. Francis Grasso is the DJ most closely associated with advancing the art of slip cueing, inspiring a revolution in the way that DJs mix.

Slip cueing gives you more direct and accurate control over what you're cueing up. This is essential for beat matching, cutting, and flying in elements from one record (like horns, vocals, pads, sound effects, etc.) over another record. It can also give you added precision in your basic transitions, useful when starting a track right on the beat. Getting comfortable with slip cueing is also a prerequisite for scratching, and facilitates techniques like extending breaks, beat juggling, and drum scratching.

Slip Mats

The first thing you'll need is a slip mat, which decouples the record from the turntable's platter, allowing you to hold the record still while the platter spins below. If you purchased

your turntable recently, it's likely that it came with a slip mat. If not, you can buy one at most any record store that caters to DJs. Some DJs make additional plastic slip mats to place underneath the traditional slip mat to give additional slippage, especially handy when scratching. The plastic sheath from a 12-inch record can be used for this purpose. Thud Rumble has come up with a hybrid slip mat called the "Butter Rug," which combines a felt-like top with a plastic-like bottom, eliminating the need to use two slip mats (Figure 15.27).

Fig. 15.27. Butter Rug slip mats feature an extra slick bottom, which emulates the properties of an additional plastic slip mat.

If your turntable came with a "grip mat," a rubber-like mat that fits on the platter in order to increase friction between the record and the platter, take it off and place a slip mat directly on the metal top of the platter instead.

Develop a Light Touch

Before you slip cue your first record, start to develop a light touch. Keep your hands totally relaxed as you perform the following on both turntables:

1. Place a record on top of the slip mat on a spinning platter.

2. Press down lightly on top of the record with your fingertips.

3. Gradually increase pressure until the record stops, but the platter keeps spinning.

Notice how little pressure this takes. Any more downward pressure is wasted energy, and will actually be detrimental to your performance. Practice repeating the above steps until you can start and stop the record at will, instantly applying the exact amount of downward pressure needed, and not one ounce more. Your goal is accuracy, agility, and keeping as light a touch as possible.

Slip Cueing Vinyl

Next, we'll practice slip cueing a vinyl record. Set up your mixer so that the channel you'll be cueing is not coming out of the main outputs, but is being sent to your headphones. Put the record on the platter, let it spin and place the stylus at the beginning of the record. As soon as you hear the first sound through the headphones:

1. Stop the record with your hand, letting the platter continue to spin (Figure 15.28).

2. Rotate the record backwards, listening as you back up to the top of the track.

3. Notice the position of the label as it corresponds to the first sound on the track.

4. Move the record back and forth over the first sound a few times (thunk-a, thunk-a), stopping at the point just before the first sound.

5. As you hold the record still with one hand, adjust the channel fader and crossfader to send the record's output to the mixer's main outputs (Figure 15.29).

6. Release the record with a slight forward motion.

Fig. 15.28.

Fig. 15.29.

Listen to the first sound the record makes, making sure it sounds natural. If the record is not all the way up to speed, or if it is momentarily too fast, adjust the slight forward motion you are using to put the record in motion.

Practice this on both turntables, until it becomes easy to slip cue, hold the record and adjust the mixer at the same time.

Next, practice going back and forth between two records, keeping track of what you're sending to the main speakers and what you're sending to the headphones at all times. Transitions, beat matching, and all other mixing techniques are predicated on your ability to handle the records, and control the main mix and headphone mix with ease.

 ## Slip Cueing CDs

You can slip cue CDs in much the same way as you slip cue vinyl, thanks to the jog/shuttle wheels and/or platter emulators built into most DJ CD players. If you learned on vinyl, this is an especially welcome feature.

If you are using a Pioneer CDJ-1000 or a Numark Axis 9 deck or something similar, locate the playback position marker on the display, which will be your visual cue instead of the label. You'll want to be in Vinyl mode on the CDJ with Auto Cue off, or in Scratch mode on the Axis 9.

If you're using a Numark CDX you can use the label to get your visual cues, similar to cueing a vinyl record. Unlike vinyl, you can rotate the marker on the metal label to 12 o'clock, then select a track using the track selection dial. The beginning of the track will then be located at 12 o'clock on the metal label (Figure 15.30).

Fig. 15.30. Setting the beginning of a track to 12 o'clock on the metal label's marker on the CDX.

Track Three

Let's cue up track three of the CD that came with the book, or any other CD with a strong downbeat. Once again, adjust your mixer to send the output of the deck you are cueing to the headphones, but not to the main outputs. Load in the CD, then:

1. Advance through the CD using the track search control(s) until you get to the track you want to cue up. If CD playback does not start automatically, press Play.

2. Press your hand down on the top of the platter to stop playback (Figure 15.31).

Fig. 15.31.

Fig. 15.32.

Fig. 15.33.

3.   Drag the platter backwards (counter-clockwise) until you get to the point where there is no sound. Notice the marker's position in the visual display (Figure 15.32).

4.   Move the platter back and forth over the first sound a few times (thunk-a, thunk-a), stopping at the point just before the first sound.

5.   As you hold the platter still with one hand, adjust the mixer to send the CDs signal to the mixer's main outputs (Figure 15.33).

6.   Release the platter to play the CD.

If you prefer, you can set a cue point before releasing the platter so you can return to the top of the track later, or you can press Pause if you need to use your hand for something else before you play the track. Most professional DJ CD decks have instant start-up when hitting Play or a Hot Cue button, which means you may be able to start a track just as quickly with one of these buttons as you can by releasing the platter or shuttle wheel. You should get comfortable doing both.

Practice cueing on both CD decks, until it becomes easy to slip cue using both the platter and the cue buttons, and adjust the mixer at the same time. Next, practice going back and forth between two CDs, keeping track of what you're sending to the main speakers and what you're sending to the headphones at all times. Transitions, beat matching, and all other mixing techniques are predicated on your ability to handle the CD decks, and control the main mix and headphone mix with ease.

Beat Matching

Beat matching is the practice of synchronizing (or syncing) the beats of two separate records to play together at the same time. This may be done in order to smoothly segue from one record to the next, or to combine elements of both records together into a live remix, or mash-up. Many dance records incorporate extended intros and outros that contain a deep groove with little or no harmonic information so that two tracks can be beat-matched and overlapped without having to worry about whether they are in the same (or complementary) keys.

Beat matching is a staple in club mixing and raves. On the other hand, some mobile DJs who concentrate on rapidly filling popular requests, rarely (if ever) find the need to beat match. Beat matching is a basic skill for the turntablist, and is a prerequisite to extending breaks and beat juggling.

If you are new to beat matching, be forewarned that it can be a bit frustrating at first. However, with a little practice, beat matching will get a lot easier, and eventually it will become second nature.

Phrasing

DJs who are especially accomplished at active mixing are skilled at phrasing. The first step to honing your phrasing ability is a thorough understanding of musical phrases.

Most dance tracks are organized in 2-, 4-, 8-, or 16-bar phrases. Syncing the beats of the tracks without syncing their phrases will result in musically unstable mixes.

Many pop songs and dance records share the tendency to use musical phrases that are multiples of four bars long. Try analyzing the records in your collection and you'll definitely find this to be true.

Think of a phrase as a musical sentence. In a popular song, an individual verse or chorus usually consists of two to four musical phrases. It might be helpful to think of this as a musical paragraph, while the song is the entire story. Just as it is confusing to interrupt one thought mid-sentence with another, it is not very cogent to interrupt a musical phrase partway through, unless you're after a particular dramatic effect.

Many dance records are more linear; eschewing typical song from elements like verses and choruses for phrases that build up to a climax over several minutes. Learn to listen for the larger, often 16-bar phrases that give these records their forward motion.

Masterful DJs will rearrange the form of a record at will by using two copies of the same record.

"I'll mix one record 16 bars beyond the other one," reveals Paul Oakenfold. "Bring it in, wash it in, wash it out, and stay 16 bars behind. If you wanted to take it even further, play the first record 16 bars, then mix in the second record, that would make it 32 bars, and you could take the needle off the first record and repeat it again."

Work toward phrasing musically when mixing and beat matching.

Syncing Two of the Same Records: Vinyl

One preliminary skill on the road to beat matching is getting two copies of the same record to play in sync with each other. You'll basically be using your slip cueing techniques to line up the beats on both turntable one and two (Figure 15.34).

Fig. 15.34. We'll call the record on your left turntable number 1, and the record on your right turntable number 2, which may also correspond to the channel numbers on your mixer.

Fig. 15.35.

To beat match two copies of the same record, cue up the downbeat of both records:

1. Let record number 1 play though the main speakers.

2. Send record number 2 to your headphones only.

3. Holding record number 2 still with your fingertips, press its Start/Stop button and let the platter spin underneath (Figure 15.35).

4. Listen to the beat of record number 1. Identify the bass drum (more on this later) and count: one, two, three, four, one, two, three, four …

5. Just before beat one comes back around, let record number 2 spin with a slight forward motion.

Listen to both records playing together. Are they playing perfectly in sync with each other? If so, slowly bring up record number 2 in the main speakers, blend it with record number 1, and pat yourself on the back.

If not, listen to the bass drums of both records (number 1 through the main speakers, number 2 in one ear of your headphones) and determine whether record number 2 is ahead or behind. Check your theory by listening to the snare hits of each record as well.

Nudging Forward

If record number 2 is behind, you can nudge it forward by using one of the following techniques (Figures 15.36–15.44).

Fig. 15.36, Fig. 15.37, and Fig. 15.38. Momentarily speed up the record using the pitch adjustment control.

Fig. 15.39, Fig. 15.40, and Fig. 15.41. Spin the label with your finger slightly faster than the record is spinning on its own.

Fig. 15.42, Fig.15.43, and Fig. 15.44. Give the spindle a few spins between your thumb and index finger, slightly faster than the platter is spinning on its own.

Nudging Back

If record number 2 is ahead, you can nudge it back by using one of the following techniques (Figures 15.45–15.53).

Fig. 15.45, Fig. 15.46, and Fig. 15.47. Momentarily slow down the record using the pitch adjustment control.

Fig. 15.48, Fig. 15.49, and Fig. 15.50. Momentarily drag the side of the platter with your finger.

Fig. 15.51, Fig. 15.52, and Fig. 15.53. Momentarily tweak (squeeze) the spindle.

Once you get record number 2 perfectly in sync with record number 1, bring its signal up in the main speakers and blend the two records together.

Fade out record number 1 in the main speakers, and assign it to the headphones only. Take record number 2 out of the headphones. Cue record number 1 back up to the top, and synchronize it to record number 2 using the techniques we just covered. Go back and forth until you can get the two records in sync quickly and easily.

Fig. 15.54. We'll call the deck on your left CD number 1, and the deck on your right CD number 2, which may also correspond to the channel numbers on your mixer.

Syncing Two of the Same Records: CDs

One preliminary skill on the road to conquering beat matching is getting two copies of the same CD to play in sync with each other. You'll basically be using your slip cueing techniques to line up the beats on both CD decks 1 and 2. If you haven't already done so, burn two copies of the CD that came with the book in order to have identical copies in both of your decks.

 Track One

Cue up track one of two identical copies of the accompanying CD on both your decks.

1. Let CD number 1 play through the main speakers.

2. Send CD number 2 to your headphones only.

3. Listen to the beat of CD 1. Identify the bass drum (more on this later) and count: one, two, three, four, one, two, three, four …

4. Just as beat one comes back around, start CD 2 right on the down beat.

Listen to both CDs playing together. Are they playing perfectly in sync with each other? If so, slowly bring up CD number 2 in the main speakers, blend it with CD number 1 and pat yourself on the back.

If not, listen to the bass drums of both CDs (number 1 through the main speakers, number 2 in one ear of your headphones) and determine whether CD number 2 is ahead or behind. Check your theory by comparing the snare hits of each CD as well.

If CD number 2 is behind, you can nudge it forward by using one of the following techniques (Figures 15.55–15.60).

Nudging Forward:

Fig. 15.55, Fig. 15.56, and Fig. 15.57. Momentarily speed up the CD using the pitch adjustment control.

Fig. 15.58, Fig. 15.59, and Fig. 15.60. Spin the platter/jog wheel clockwise with your finger. Be sure you're in CD (not vinyl or scratch) mode. This will speed up playback momentarily.

Nudging Back:

If CD number 2 is ahead, you can nudge it back by using one of the following techniques (Figures 15.61–15.66).

Fig. 15.61, Fig. 15.62, and Fig. 15.63. Momentarily slow down the CD using the pitch adjustment control.

Fig. 15.64, Fig. 15.65, and Fig. 15.66. Spin the platter/jog wheel counter-clockwise with your finger. Be sure you're in CD (not vinyl or scratch) mode. This will slow down playback momentarily.

Once you get CD number 2 perfectly in sync with CD number 1, bring its signal up in the main speakers and blend the two CDs together.

Fade out CD number 1 in the main speakers, and assign it to the headphones only. Take CD number 2 out of the headphones. Cue CD number 1 back up to the top, and synchronize it to CD number 2 using the techniques we just covered. Go back and forth until you can get the two CDs in sync instantly.

Tips and Tricks

- You can get a phasing or flanging sound by having two records playing *almost* in sync. Try it.
- Some mixers' cue sections let you split what you are hearing on deck number 1 to the left side of your headphones, and deck number 2 to the right side of your headphones (or vice versa). This can sometimes be helpful when determining whether the track you are trying to sync is ahead or behind.

Beat Matching Two Different Records

To beat match well, you'll want to collect records that will potentially work well together. This usually means the difference in the record's tempos, expressed in beats per minute or BPM, is close enough to bridge using your decks' pitch controls, without making the track sound ridiculous (unless that is your goal). Selection is key, and coming up with a pair of records that are incredible together and combining them in an innovative and effective way is a great feeling.

Many genres of dance music are at least partially defined by their tempos, making it easier to find tracks within the same genre that will be close enough to beat match.

The following is a rough list of tempos for a few different genres:

Style	Approximate Tempo (in bpm)
Reggae	70–82
Trip Hop	80–92
Hip-hop	89–105
Acid Jazz	110
House	119–128
Break beat & electro	130–132
Progressive	130–135
Trance	135–140
Techno	140 and up
Hardcore	150–160
Drum 'n' Bass	160–170

You can calculate BPM with a stopwatch, or any clock that shows seconds ticking by. The simplest tempo to calculate is 60 bpm, which equals one beat per second and is considered a very slow tempo. Another tempo that's easy to find using a clock is 120 bpm (two beats per second), which is generally considered medium to up-tempo.

In order to calculate the BPM of a record, simply do the following:

1. Count the beats that occur during 15 seconds.

2. Multiply by four.

For example, if you count 35 beats in 15 seconds, the BPM is 140; if you only count 27 beats in 15 seconds, the tempo would be 108.

The basic unit of measurement when determining tempo is the quarter note. Written music will most often specify tempo with the phrase: *quarter note =*, followed by the number of beats per minute, for example: *quarter note = 120* or *quarter note = 97*.

Automatic beat detection technology has seen significant advances over the last few years. The "Beat Keeper" system by Numark is now built into many of their CD players, mixers, and even the TTX turntable. Beat detection technology works by analyzing repeated low-frequency information (kick drum), and mid/high-frequency bursts (snare), to extrapolate the drum pattern and its tempo. It works best on tracks with a 4/4 time signature and a steady (repetitious) beat, especially dance music that employs a "four on the floor" kick drum pattern, more on this later. Some systems allow you to tap in the beat if the computer has miscalculated.

Tips for Getting Organized

A few lucky DJs have a photographic memory—for the rest of us, we need to develop a system:

- Determine the BPM of the records in your collection. If you haven't done this up until now, go back and BPM the tracks that you already use.

- BPM new records as you acquire them.

- Keep a master list.

- Write BPM's on your album jackets.

- Organize your records according to BPM.

- Put the BPM in the title of each tune in your digital music database.

A few lucky DJs have a photographic memory—for the rest of us, we need to develop a system.

Music Notation

In order to fully grasp what goes into a drum pattern, how to nail its BPM and blend it with other patterns, it's extremely helpful to understand note values and time signatures. If your eyes glazed over in music class and you decided that reading music was too difficult, don't panic! You'll be pleasantly surprised at how easy it is when you're using it to analyze beats.

You might want to fold down the corner of this page so you can refer back to it as you are checking out the following drum grooves. Trust me, compared to splitting your brain, this is a piece of cake! See Figure 15.67–15.75.

Fig. 15.67. Music is sliced up into measures or bars by bar lines on a staff.

Fig. 15.68. Most bars are four beats long, and quarter notes each get one beat. This information makes up the time signature, the most common of which is 4/4.

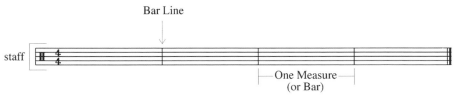

Fig. 15.69. Notes are made up of heads, stems, flags, and beams. Whole notes, half notes, quarter notes, eighth notes and sixteenth notes each get progressively more ink.

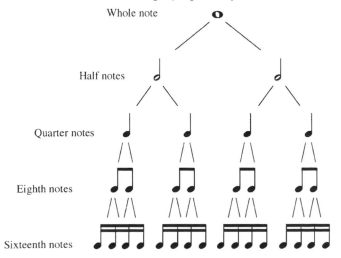

Fig. 15.70. In 4/4 time, a whole note gets all four beats (a whole bar), half notes each get two beats (half a bar), quarters each get one beat (a quarter of a bar).

Fig. 15.71. In 4/4 time, eighth notes get half a count, so there are two eighth notes for each beat (eight for a whole bar—thus the name).

Fig. 15.72. Sixteenth notes happen even faster; four per beat.

Fig. 15.73. Triplets divide things into thirds rather than in half; eighth note triplets each get one-third of a beat.

Fig. 15.74. Other than 4/4, the most common alternate time signatures are 3/4 (three beats per measure), and 6/8 (six beats per measure where the eighth note gets one beat).

Fig. 15.75. The staff provides lines and spaces to write different pitches, or to represent various parts of a drum track.

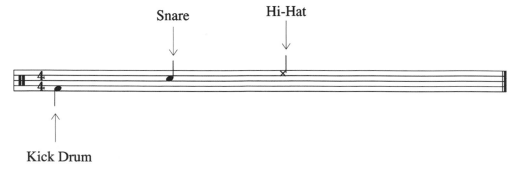

Components of a Drum Track

Some records are easier to BPM than others. Let's examine some widely used beats and their components. Listen to the beats on the included CD, and practice determining their BPM. You can check your accuracy by referring to the track listing in the back of the book.

A lot of dance music features what's called *four on the floor*, which means that the bass drum (also called the kick drum) is playing quarter notes in 4/4 time.

While four on the floor is common in most genres derived from house and techno, it is far from new. In the early days of swing, drummers often laid down four on the floor to keep the dancers moving. Field recordings of African and Native American tribal music suggest that a strong quarter note pulse in 4/4 is something that may well be woven into the human psyche.

On dance records, the bass drum is usually the lowest pitched and loudest percussive element. It may be a recording of an actual kick drum, or the electronic equivalent, featuring multiple samples triggered together to lay down a solid boom.

The other essential factor in determining BPM is the snare, or its electronic equivalent, which could be handclaps or a burst of white noise. The snare sound is almost always higher in pitch than the kick sound.

Most often, the snare is placed on the upbeats, which in 4/4 time is beats 2 and 4. When combined with a kick drum laying down four on the floor, the resulting pattern will look like this (Figure 15.76).

Fig. 15.76. A typical four on the floor pattern with just a kick and snare.

 Track Thirty Seven

Notice that both the kick and the snare are playing together on the upbeats (2 and 4), while the kick is handling the downbeats (1 and 3) on its own.

The third factor present in most drum tracks is the high-hat or its equivalent, which could be a shaker, a tambourine, or practically any high pitched, short sample. The function of the high-hat is usually to subdivide the beat. Often the high-hat sound will be playing eighth notes along with the quarter notes of the kick and snare (Figure 15.77).

Fig. 15.77. The high-hat playing eighth notes over the four on the floor beat.

Track Thirty-Eight

On rock-oriented tracks, often the kick lays out on beats 2 and 4, concentrating on the downbeats and leaving the upbeats for the snare (Figure 15.78).

Fig. 15.78. A basic rock-oriented beat.

Track Thirty-Nine

Sometimes the kick will add in attacks on the eighth note immediately before or after a downbeat as in the following example (Figure 15.79).

Fig. 15.79. Another rock-oriented beat, with a slightly busier kick pattern.

Track Forty

What is often referred to as a disco beat features 16th notes in the high-hat along with a four on the floor kick and the snare on the upbeats (Figure 15.80).

Fig. 15.80. A basic disco beat.

Track Forty-One

Hip-hop beats add swing into the equation. Instead of the high-hat keeping even (or straight) 8th or 16th notes, they are swung, meaning that they have more of a triplet feel (Figure 15.81).

Fig. 15.81. A Hip-hop beat.

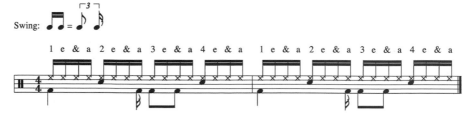

Track Forty-Two

Some slower tunes are in 6/8 rather than 4/4. A couple of examples are "It's a Man's World" by James Brown and "Oh Darling" by the Beatles (Figure 15.82).

Fig. 15.82. A beat in 6/8.

Track Forty-Three

Most jungle and drum 'n' bass records have a beat based upon a sped up version of a break known as the "Amen beat." It looks a bit more complicated, and contains quite a bit of syncopation (Figure 15.83).

Fig. 15.83. A jungle or drum 'n' bass beat.

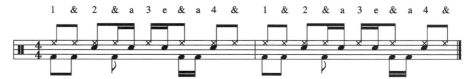

Track Forty-Four

So how will knowing all this help you? When beat matching, you're lining up the bass drums and snares so they work together, rather than in conflict with each other. Knowing the function and tendencies of the kick, snare, and high-hat makes it easy to figure the BPM, sort out other potential challenges, and program your own beats.

Tracks to Beat Match

I've included a couple of tracks on the CD that came with this book that should beat match nicely, and work together harmonically.

Tracks One and Two

On the enclosed CD, try matching tracks one and two.

In order to beat match track one with track two on two different CD decks, rip and burn two identical copies of the CD.

If you're mixing with Final Scratch or another laptop interface, you can rip the two tracks from the CD into your audio bin.

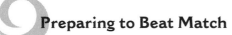

Preparing to Beat Match

As we move ahead, you'll notice there are many options and judgment calls to be made. Don't let this throw you; you'll be developing your taste (and consequentially your style) as you experiment. Take a deep breath, clear your head, and *listen*.

The following will help ease you into beat matching tracks with different tempos.

First, determine the tempo (BPM) of both tracks.

Second, answer these questions:

• Which track is faster?

• How much (in BPM) will the slower track need to be sped up?

• When you speed it up that much, how does it sound?

• How does the faster track sound slowed down?

• Does it sound better to meet somewhere in between?

• Do the tracks work together harmonically, or musically?

Third, listen to the phrasing of both tracks.

Once you've considered the above, beat match the two tracks by matching their BPM with your deck's pitch controls, then synchronizing their beats with the skills you mastered in the previous section on syncing two copies of the same beat.

Things to Consider

Sometimes slowing tracks down can make them seem bigger, but if done too much it may deflate the energy of the original. Speeding a track up a bit tends to give it more edge, although it's possible to go too far and make things seem frantic. Don't be afraid to try things, even radical things, but be honest with yourself about the results.

Keep in mind that pioneering DJ David Mancuso wouldn't use the pitch control at all, choosing to honor the recording artist's original intentions. On the other hand, genres such as jungle and drum 'n' bass are based upon radically speeding up an old funk beat.

Harmonic Mixing

Harmony, the chordal and melodic instrumental and vocal parts of each record are extremely important to this process. If you beat match and mash up two records that are in two different keys, they have the potential to clash horribly. They also have the potential to blend beautifully, if they are in the same key, or in relative keys.

Each major key has a relative minor key, and vice versa. The relative minor key is located a minor third below the major key center; for example, A minor is the relative minor to C major.

When moving from one record to the next, it is generally most pleasing to mix into a relative key, or a key that is a perfect fourth or a perfect fifth away from the key of the original record.

You can determine the key of a record by finding the tonal center on a piano or keyboard synthesizer, or by checking on one of the available lists, like those available at harmonic-mixing.com. Mark Davis of harmonic-mixing.com has created something called the Camelot Easymix System, which uses a wheel to simplify harmonic mixing (Figure 15.84).

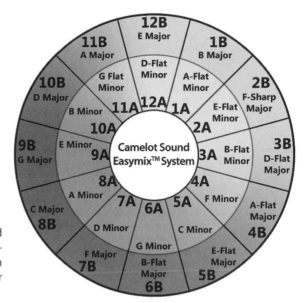

Fig. 15.84. The Camelot Easymix wheel can be viewed as a clock, with musical keys seen as "hours." To use harmonic mixing in your DJ sets, you can transition between songs by subtracting one hour (−1), adding one hour (+1), or staying in the same hour as your last song.

In recent years, digital technology has made it possible to change the key of a record independent of the record's tempo (Figure 15.85).

With CD players like the CDX and turntables like the TTXs which offer a Key Lock feature, you can now adjust the pitch to find a complementary key, lock that key into place using the key lock feature, then go off searching for the tempo independent of pitch.

Fig. 15.85. The *Key Lock* control on the TTX.

There are many computer programs that do this now as well, including some that claim to do this automatically, by having a database of keys built in. While this may seem akin to magic, there is a limit to how far from the original pitch you can take a track before digital artifacts begin to appear.

Making Mix Tapes and CDs

Making mix tapes, even if just for your friends at first, is a good way of taking your practice sessions to the next level. Even though almost everyone now uses CDs, many people still call the result of recording a live mix a "mix tape."

Recording your practice sessions keeps you from just brushing off your mistakes; since someone is potentially going to hear it, you pay attention more and make extra effort to turn in a better mix. It's just human nature.

Record your practice sessions through the stereo *record outputs* of your mixer so you can scrutinize your transitions later, while you're not splitting your brain.

As you get better, your mix tapes can become your calling card. There are countless examples of DJs, reputations preceding them because somehow, someone got a hold of their mix tape. We'll visit this more in next chapter.

16

Set Building Skills

When you first start mixing, your goal is simply to mix two records together smoothly. That goal will take many tries and re-tries. Once you have developed basic beat matching and mixing skills with various pairs of records, it will be time to create a set.

What is a set? A set is a group of records, spun one after the other (sometimes moving back and forth in your playlist with bits of previous or future records), to create a complete musical presentation. A set can be 30 minutes long or 14 hours long, and it can be spun for no one or for an audience of thousands.

"You have to feed off the crowd … you have to play with them," says House DJ Keenan Orr. "I love to surprise the people that I'm playing for, and I try to get them to scream at least one time."

Qualities of a Set

Think of some recent or memorable DJ sets you've heard. Choose a set you experienced that really stands out in your memory. What made the set so good?

Some things that people (and promoters) notice are:

- Mixing skill
- Track selection
- Programming
- Originality.

Mixing skill: Mixing skill refers to your ability to use the equipment well to make flawless, interesting, and creative mixes with your records. Mixing is your chance to really make the music your own by incorporating your choice of blends, tricks, scratches, or effects. Your choices will also distinguish you from all the other DJs out there, so you'll want to develop a few creative methods that are yours and yours alone.

Track selection: Track selection refers to the quality and appeal of the tracks you choose to play. Choosing the appropriate blend of popular or accessible tracks along with new, unique tracks that go well together is part of good track selection.

Programming: Programming refers to the flow of your set. Does your set go anywhere? Does the audience feel like the set progresses smoothly toward some goal? Are they exhausted 20 minutes into the set, or do they appear uninterested in what you are playing? Below we'll list some factors that will affect the flow of your set. In addition, you will need to be able to modify the flow of your set based on the reaction of your audience on that particular night and in that particular venue.

Originality: You've heard it by now: everyone's a DJ. In many cities, there are simply too many DJs and too few listeners. In order to succeed as a DJ today, you will have to distinguish yourself. You can distinguish yourself by developing a signature sound, employing signature tricks and effects, developing a signature track selection, a look, or anything that makes you stand out above the rest. Perhaps you wish to create a DJ persona, complete with props and costumes. Perhaps your uniqueness will be in your consistent ability to rock an exhausted crowd at 8:00 AM. Maybe your CDs will be a favorite in your city because you incorporate local themes into the music.

"DJ sets are all about NOW. It's about here and now," according to techno producer Jon Rivers. "It's not about last night, or where you played last week, or what you were thinking about on your way here. The best DJ sets I've heard reflect the exact moment in time when they occur, between the DJ and the audience, creating that elusive goal: the VIBE."

Factors Affecting Your Set

As a DJ, you will choose the mood, pace, length, and content of your set. If you are spinning for yourself in your room, you can play whatever you desire at that moment.

Think again of that excellent set you heard. In addition to the intrinsic qualities of the set itself, there probably were other factors that influenced how you and others felt about it. These factors might have included:

- Your personal mood and energy level.
- The time of the day or night that the set was played.
- The size of the venue and of the crowd.
- The type of people in the crowd.
- The age of the crowd.
- The geographic location of the venue.
- The theme of the night or of the room.
- The skill of the previous DJ.
- The mood of the crowd based on current events, the economy, the weather, etc.

If you are playing for an audience, you will want to take at least some of the above factors into account when planning for and performing your set. Your audience has needs, and you'll probably want to fulfill at least some of those needs.

Some DJs choose to "educate" their audience and play only what they want to play. That approach can work well if the musical style is a popular one that generally pleases audiences, or if the DJ is well-known enough to draw an audience interested in experiencing that style. However, if the musical style is very esoteric, the DJ may find the audience wandering away if they don't hear anything that catches their ear.

On the other hand, playing nothing but requests or "anthems" you know are popular can leave you feeling like a "human jukebox" and can sound unoriginal to your listeners. DJs are artists and musicians, and it's important that whatever you play makes YOU feel good. Your personal style will show in the tracks you choose and how you mix them together. How you feel about the records you play will come through your set to the audience. If the tracks move you and you let it show, chances are your audience will find your enthusiasm contagious. If, however, you are bored with what you play, they will most likely sense that as well.

One approach is striking a balance between playing tracks you know are crowd pleasers, and tracks that move you personally but might be new to your audience. It's a very fine line to walk, and it's a personal decision you will need to make as to how much to cater to your audience. Of course, in the end, whether or not your sets fill the dance floor will determine whether or not you get invited back for another gig. Time and practice helps most DJs learn how to respond to the factors above, play what they want, and keep their audience happy, all at the same time.

Developing Your Personal Style

Some new DJs attempt to emulate their favorite DJs when they start to play. It is only natural that you will initially be drawn to the same tracks as the DJs that helped you make the decision to become a DJ. Some new DJs even try to replicate exact sets performed by their favorite DJs.

Having a DJ to mimic can be very useful at first. The amount and variety of music available can be overwhelming, and following the tastes of your favorite DJs can help narrow the field of music and give you a framework within which to spin. However, as soon as possible, it's important to let your own style shine through. Be sure to keep your mind open to labels, tracks, and artists NOT spun by your favorite DJs. Simply listen to a few new records each time you go to record shop, and if something catches your ear, buy it. Over time, those tracks that you alone choose will shape your personal style. In general, people will be far more interested in hearing your personal style than hearing a set they heard previously spun by your favorite DJ.

Planned Sets versus "On the Fly"

You may want to plan your first few sets out. It can be overwhelming to switch from spinning in your bedroom to spinning at for an audience. Here are a few reasons:

1. The gear may be different and unfamiliar to you.

2. It may seem "so loud you can't hear anything."

3. People will be asking questions, making requests, and otherwise bothering you.

4. There will be other DJs in the booth looking over your shoulder.

5. You may have stage fright.

6. You may get beer spilled on you at any time.

In other words, in addition to dealing with all of the above, you may not want to also have to choose amazing records one after the other for your set. Planning your first few sets is understandable.

However, planning restricts you in many ways. You cannot respond to the audience, the night, the venue, or do anything that is unique to the moment with a planned set. Also, if you make a mistake, you may not be comfortable enough with spinning your records in many different ways, so you may panic not knowing what might work next.

It's risky to play a set "on the fly," because there is the possibility that you will choose a record that doesn't go so well with another, or that you will make a mistake. The risk is well worth it, however, when you pick a record that you think might go together with another and the combination turns out to be amazing. There is no greater feeling than performing that serendipitous, perfect mix for an appreciative audience.

Chunking

A useful compromise is to practice with your records enough that you know certain "chunks" of 2–3 tracks that go well together. You can choose and arrange your chunks depending on how the night is going and all the factors discussed earlier. Chunking gives you flexibility without requiring that you memorize every combination of every record that you own.

Hip-hop promoter and aficionado LaTina "Mother Earth" Mobley says that what separates talented DJs from the others is, "The selection of music, the way they blend the music. They don't start their set off with the most popular thing, it goes up from low to high then it goes back down. I like it when DJs go from old school to new school and even back again."

Demo CDs versus Live Sets

Demo CDs (formerly known as "mix tapes") are your opportunity to show an audience your very best work; your ability to craft the best mixes and choose the best records to express a certain theme or idea. For these reasons, many DJs choose to carefully plan their CD sets and spin them again and again until they are as flawlessly mixed as possible. Other DJs, however, believe that their style is best reflected in unplanned sets, spun on the fly and recorded. These DJs believe that spinning a set over and over takes the "freshness" out of spinning a set, so they choose to spin and record a different set each time, until they create something with which they are satisfied.

However you choose to make your demo CD set, be sure it is as flawless as possible. If you are just handing them out to friends, it may not matter if there are a few mistakes. However, if you'd like to perform, there are simply too many DJs today to expect that promoters will choose a CD with train wrecks on it over one spun flawlessly. If a DJ can't spin well in the comfort and peace of their own home, why should a promoter believe they will do well with the pressure of an audience, unfamiliar gear, and much bigger sound?

Some DJs wonder about editing out train wrecks and other mistakes in their mixes, using computers and software. It is indeed possible to remove many mistakes using these programs. However, for new DJs, the process of developing and perfecting your demo CD is part of the training process needed to prepare for performing. If you "cheat," and edit your mix, you are not offering the promoter an accurate representation of your mixing ability. Since performing is far more challenging than spinning alone at home, you can expect to make many more mistakes live if you have not perfected your skills at home.

Once DJs have established themselves as solid performers, some choose to make "studio mixes" that involve much editing and enhancing of the mixes through a variety of processes. At this point, it is not cheating to use this approach because the DJ has already proven his or her ability to perform live, and is actually employing a new set of skills designed to take their mix CDs to the next level.

Bad Sets

Every set will not be perfect, or even acceptable. Many DJs have significant stage fright due to the fear of delivering a less-than-stellar performance. It's true that standards have become quite high for DJs, but those standards are mostly within the DJ community and not as much among audiences. Audiences have become more savvy, and they do hear and respond loudly to outright train wrecks. However it's the other DJs you know, and you your-self, that will notice every little tiny mistake, slipped beat, or botched harmony in your set.

One of the best ways to overcome stage fright is to have an awful night and survive. Many DJs fear that they will be laughed at, or never called for another gig, or shunned by the DJ community if they deliver a bad set. The reality is that everyone has a bad night now and then—even the DJs that have been spinning for 20 years or more. As long as you *generally* deliver quality sets, you will be completely forgiven for the occasional bad night. Try to learn from the experience … should you have practiced more? Did you play tracks you hadn't yet mastered? Was there a piece of gear you need to get more familiar with? Did you have one beer too many? As long as you learn from your mistakes, your sets will get better and your bad nights will be fewer and further between.

The Zone

Your sets will ultimately define who you are as a DJ. They are the final result of your practic-ing, your musical selection, your skills, and your personal style. Rather than get too caught up in the technical details of your set, try to let your musical "soul" shine through. Some DJs refer to spinning a good set as "getting in the zone." Once you can let go and stop over-thinking every detail, you know you're on your way to delivering unique and crowd-pleasing sets. Practicing often and with many different records will free you up to try new things and to take risks when you perform. You will also find yourself more open to interacting with your crowd, responding to the unique factors of the night, and worrying less about small mistakes in your performance. Taking risks and making yourself somewhat vulnerable are, as with many musical pursuits, essential to making the leap from bedroom hobbyist to a seasoned performer that people will line up to see again and again.

Mobile DJ Skills

With the possible exception of "bedroom DJs" (people who mix and scratch for fun at home), there are probably more mobile Disc Jockeys out there than any other category of DJ, and for good reason. You can make some good dough.

High school and middle school dances and proms, wedding receptions, sweet 16 parties, bar and bat-mitzvahs, anniversaries, reunions, birthdays, corporate events … every week there are tens of thousands of functions played by mobile DJs. DJing the events listed above is definitely a "service industry," and it's important to check your ego at the door.

"It's not really about the DJ at a sweet 16 party," advises Eric Sans, AKA Kid Koala, who played mobile DJ gigs as a teenager.

In many ways, being a mobile DJ is like playing in a Top 40 band. You are there to spin the hits and keep everyone happy.

Many current stars of the DJ world started out playing modest mobile gigs, from Mixmaster Mike in the garage parties of Dailey City, to DJ Kool Herc playing through a guitar amplifier in the recreation room of his sister's building in the South Bronx.

Even if you aspire to international DJ stardom, local mobile DJing can be a good way to learn your craft, assemble some equipment, and earn some money. Pay can be less than $50 for a neighborhood kid starting out with a small thrown-together system, to over $5000 for a seasoned professional with a colossal PA and intelligent lighting system working a large corporate party.

Chris Roman has been a successful Mobile DJ for over 20 years, starting his own mobile DJ service with a friend when he was 15 years old. Within a year, they were running a company that covered up to 10 gigs per weekend. Chris did quite a bit of club work along the way, but has decided to stick with the mobile gig for reasons he'll explain in this chapter.

Chris now divides his time between mobile DJing and developing products for Numark. I asked Chris to share some of his insights.

What does it take to compete as a mobile DJ in terms of investment?

The mobile DJ brings his own equipment. The investment is big and the time you work is doubled. For a four-hour gig you might spend 12 hours with loading in/out on both ends and setting up. For a basic professional audio set-up, the DJ will spend $3000 to $10,000 and with lighting you can figure on that much again.

My mobile rig varies by event but for larger shows, 1–2000 attendees, I bring in the intelligent lighting and massive sound systems. More recently I've gotten into video dance parties. It's amazing how this is coming back; at this point I'm doing video about half the time. I have two video projectors, and I just brought a third screen. I've been using a tripod screen with a skirt, and sometimes I also use two 149 inches inflatable screens designed for outdoor theater, which I bought for $150 each. They take up no space in my truck, but they're not exactly the kind of thing you bring to a prom.

I also just purchased two rear projection screens off of eBay. They're big: 10 feet × 10 feet! Rear projection can be crisper, and the screens can be up where you are.

I used to work off of CDs, and my collection was very expensive. Mobile guys adopted CDs early because the space and weight required is 1/10th that of vinyl. I remember the days of taking six crates of records, with 300 or so 12 inches singles and LPs and a couple of tool boxes full of seven inches singles.

Most mobile guys now are onto hard drive-based computer solutions. I use the Numark HDCD-1, which is a hard drive unit. One new problem for the professional DJ is that now the barrier for kids is so low, if they download their music illegally. I legally own the music I play, and with the larger hard drives these days, I can have everything at a less compressed size.

What mentality is required of mobile DJs?

DJing is as much about psychology as it is about musical skills.

Mobile work means you never know what you might do from night to night. Beat mixing may not be required or may even be discouraged. It's exciting because the risks of failure are high. A 16-year-old party means you better know what is hot for this group in the town you're working in. Hip-hop might not be so hot if your crowd is into rock.

Weddings bring a certain degree of formality and structure. If you know the ropes this type of event can often be easier.

Pay is based upon going rates, size of job, and reputation. It's usually much better than club work, unless you're one of the Paul Oakenfolds of the world and your name will draw a large crowd. Work in this [mobile] area requires continual legwork though, and the work comes and goes.

As a mobile DJ, what conversations do you have with prospective clients when they call to hire you?

The first question is how much. This often comes before you have the opportunity to spec the requirements to do the job correctly. The second question usually has to do with experience and referrals. My answer varies upon the needs of that client. For instance, if you are doing a sweet 16 your client will want to know if you have the music and understanding to make the 16 year old and her friends happy.

Bar mitzvahs and weddings require unique skills that only come with experience. A client will often ask about your music, experience, and how much you might talk.

After you have the gig, what sorts of things do you do to lay the groundwork with clients before the job?

Things you need to know about are crowd size and room size so you bring the right amount of equipment. You also ask about musical tastes, if lighting or additional sound systems are needed in other rooms, special presentations planned during the night, etc.

Load in requirements for bigger events are talked about with the function facility and with bigger events union labor must also be considered.

What tips do you have for setting up before a mobile DJ gig, both equipment-wise and interfacing with the client and the audience before you start?

Make sure your set-up is clean and clear of wires everywhere. Bring a list of your music so people can make requests off the list. With younger crowds this is essential to keep you sanity while 20 people are in your face all night.

Bring enough sound equipment. For the high school you can never bring too much. On the other hand, for a wedding, big speakers will often create the illusion of a big sound and the older crowd will complain about how loud it is even before you plug in your amp.

ALWAYS BRING BACKUP! In a mobile event you are the only entertainment. You cannot afford failure in any way. It would likely ruin someone's day other than your own and I've heard of guys getting sued after ruining a brides day. In this case have a backup amp and CD system in the car. Even if you have to change songs one at a time on a Walkman, it's better than nothing. I recall a night when I fried an amp and having the backup in my car actually meant a big tip at the end of the night because I was up and running again within five minutes and the client recognized my preparedness.

It's time to begin your set. No one is yet on the dance floor. How do you decide what to lead off with and why?

This is the hardest of all questions because depending upon your crowd your night could end before it begins.

For a mobile DJ, the crowd is usually instant and the first song you even play is creating a mood. In a wedding you can make formality your friend and move your key participants to the floor with a bridal party dance.

With a high school or college gig, the goal is to motivate from the time the door is opened. Don't play the hot stuff too early, unless it looks like your crowd won't even start without it. Play the "almost classics" they know for the first 20–30 minutes, in high school this is last years songs, in college this means songs from the past 5–10 years. When they are ready and you have 30–50 percent of your audience moving either on the floor or on the sides, drop in whatever might be the hottest thing out. By playing it this early you can do it again at the end of the night as well.

18

Extending Breaks

Originated by Kool Herc and perfected by Grandmaster Flash, the technique of extending breaks was first employed to give the B-boys and B-girls more of an opportunity to show off their best dance moves. It was also used to give MCs a live loop to rap over, which turned into the basic building block of Hip-hop production. The technique also forms the basis for beat juggling.

With two copies of the same record, you can extend (or repeat, or loop) any part of the record you want. With practice, you can do this seamlessly.

Before extending a break, you'll need to determine where you want to begin and where you want to end. You'll also want to mark the beginning of the break as a visual cue.

Some breaks occur in the middle of the record, others are used as intros. You can actually extend any portion of a vinyl record, provided you have two copies. Many funk records have a "breakdown section," where everything drops out except for the drums and percussion, and sometimes the bass.

One of the features of most CD turntables is their ability to loop portions of a single CD. This is a lot easier and less dangerous to pull off live, making it more reliable but potentially less exciting than extending a break off of two copies of a vinyl record. We'll look at looping a portion of a CD after we extend some vinyl breaks.

It's easiest to start off by finding two copies of a record with a rhythmic introduction of at least two to eight bars that happens before the vocal enters. Many Hip-hop and rap records will fit the bill nicely, as will classic tracks like "Low Down" by Boz Skaggs, which Grandmaster Flash pioneered the use of for this purpose.

Marking the Label

In order to have a visual cue to help us locate the downbeat of the track, we're going to place a mark on the record label.

You can mark the label to locate the beginning of a break (or any other event) by using a label meant for an audiocassette, mini-disc, or mini-DV videocassette, an adhesive dot from an office supply store, or even a Sharpie felt tip marker and a straight edge (Figures 18.1 and 18.2).

Fig. 18.1 and Fig. 18.2. Marking a record label with an adhesive dot and with a Sharpie.

Fig. 18.3. Viewing the record label as a clock.

Look at the label as though it were a clock, with 12 o'clock being straight up, 6 o'clock straight down, etc. (Figure 18.3). This is what Grandmaster Flash called the "clock theory."

In battle mode, a straight-arm TTX1 has the stylus meeting the record at approximately 1 o'clock. In standard mode with an s-style tone arm, the stylus meets the record at about 5:30.

Some DJs mark the label at 12 o'clock; others point the mark directly at the stylus (Figures 18.4 and 18.5).

Fig. 18.4 and Fig. 18.5. Marks pointing to 12 o'clock, and pointing to the stylus.

Marking the label at 12 o'clock is easy to see, which can be helpful in adverse conditions (dark DJ booths or glaring stage lights). Pointing your mark directly at the stylus gives a more literal depiction of where the cue point is, and will remain constant when used in various set-up modes and with different tone arms. Both methods are in widespread use.

Experiment to determine which method of marking works best for you by trying the techniques explained below with marks at 12 o'clock, then moving the mark to point at the stylus and executing the techniques again. Which method best helps you consistently hit the cue point?

Once you decide on a marking method, remain consistent whenever you mark your records.

Two Bar Backspin

Once you've marked both records at the downbeat of the selected track, you'll use those marks to help loop a two bar break. Next, you'll need to determine the end of the section you want to loop, and get the hang of backspinning back to the top.

Cue both records to the downbeat of your selected track, and place the crossfader in the middle.

1. Start the record on your left, counting eight beats (or two bars) as it plays.

2. Just before the ninth beat, stop the record by placing your middle finger between the label and the last track (Figure 18.6).

3. Backspin the record to the top of the track, using the mark to count the number of revolutions (Figures 18.7–18.9).

Fig. 18.6. Stopping the record.

Fig. 18.7– Fig. 18.9. Backspinning to the top of the track.

4. Let the record play again, counting eight beats (or two bars) as it plays.

5. Repeat steps 2 through 4 until you are fluid at this task.

Repeat this exercise using the record on your right.

Two-Bar Silent Backspin

The next step is to backspin without listening to the backspin, using the mark to visually count the number of revolutions, which is usually about three for a two-bar phrase.

Cue both records to the downbeat of track one. This time place the crossfader all the way to the left.

Let the left record play for two bars, stopping it with your middle finger before the ninth beat just as you did in the first two steps of the previous exercise. Then:

1. Throw the crossfader all the way over to the right, effectively muting the left turntable.

2. Backspin the left record to the top of the track, using the mark to count the number of revolutions.

3. Throw the crossfader all the way back over to the left.

4. Let the record play again, counting two bars.

5. Repeat steps 1 through 4 until you are fluid at this task.

Repeat this exercise using the right turntable, substituting the words "left" and "right" in the above examples.

Extending a Two-Bar Break

Once you have a gained fluidity with the previous tasks, you are ready to tackle extending this break *in time*.

Cue both records to the downbeat of track one, and move the crossfader all the way to the left.

1. Start the record on your left, counting the beats as it plays.

2. Press the Start button for the record on your right as you hold the record in place.

3. Just before the ninth beat, slam the crossfader all the way over to the right, releasing the right record on the beat.

4. Backspin the record on your left back to the top of the track as the right record plays eight beats.

5. Just before the ninth beat, slam the crossfader all the way over to the left, releasing the left record on the beat.

6. Backspin the record on your right back to the top of the track as the left record plays eight beats.

7. Repeat steps 3 through 6 until you get the hang of it, or until you just need to stop to keep your sanity.

Tips:

It will take some practice to coordinate these moves and keep the beat steady. Don't get frustrated.

If you backspin too far, there will be a pause before the downbeat when you let the record spin. On the other hand, if you don't backspin far enough, or are too slow with the crossfader, you may clip the downbeat of the record you are releasing.

The keys to seamless beat extension are:

1. Nailing the timing of the crossfader.

2. Nailing the timing of the record release.

3. Accurately backspinning to the top of the sample.

4. Blending all these moves into one fluid technique.

Remember, extending breaks is pointless if it doesn't *groove*. Accuracy and flow are of supreme importance.

If your needle is skipping, try lightening up your touch; first as you come in contact with the record, then as you backspin, finally as you put the record in motion. Avoid putting unnecessary downward pressure on the record, as downward pressure on one side will lift up the opposite side, making it hard for the needle to stay in the groove. It may also help to increase the weight on the tone arm, although too much weight will accelerate burn in on your records.

Once you start getting the hang of extending this two-bar break, try cutting it down to one bar, or increasing the extension to four bars.

Look for other sections of records to extend, and creative ways to work this into your sets.

19

Scratch DJ Skills

Once you're scratching, you've crossed the line from using the turntable as a playback device, to using it as a musical instrument.

Scratching can be simple or intricate. Most of the time it's brazenly percussive. Depending on the source material, scratching can accentuate a drum pattern, add rhythm to a vocal, give additional weight to a lyric, or cut through a beat in a blaze of rapid-fire, in-your-face syncopated mayhem.

Scratching is called scratching because of the way it sounds, especially when using the ever-popular "Ahhh" and "Fresh" samples. The record is not *literally* being scratched; physical scratches are not being gouged into vinyl. Rather, you're using your hands to manipulate the record back and forth in any number of musical patterns.

Record Wear

For the turntablist, the record and stylus are the two components providing the musical vibrations, like a reed for a sax player, or a bow and string for a violinist. All of these components eventually become worn and need to be replaced.

Records do wear over time, and scratching increases wear by repeating certain portions more than others, but scratching does not immediately destroy records. Many skillful DJs own at least some records they've been scratching for years. If your needle, tone arm, and counter weight are set up well, your hands are clean and you play with accuracy and a light touch, wear can be minimized. Wear from scratching is called "burn in," and it sounds like white noise. In some applications, burn in can be a desirable sound.

There are a lot of "scratch records," on the market that feature samples and sounds created for scratching, and beats that work particularly well for juggling. Many of these records are readily available, so once they're worn out, they can be easily replaced.

This is not the case with records that are out of print. When I find a clean or pristine copy of a classic vinyl record, I'll make a good digital transfer of it before playing it many times or using it to scratch with. Then I'll practice cutting and scratching with it on a CD or hard drive deck like the Numark CDX or HDX, or on a computer interface system like Final Scratch or Serato Scratch Live to save wear on the record itself.

Musical Origins

"Scratches" are to turntablists what "licks" are to guitarists. We know the origins of many guitar licks; certain blues licks are attributed to Robert Johnson and Muddy Waters, Jimi Hendrix licks are often imitated by rock guitarists, Wes Montgomery and Charlie Christian added many early jazz licks to the lexicon, Eddie Van Halen pioneered a whole category of licks known as "tapping."

We also know the origins of many scratches. Grandwizard Theodore investigated the concept of scratching as a form of musical expression in the 1970s. Many of the first scratches were pioneered and refined by Theodore and Grandmaster Flash. Philadelphia DJs Spinbad, Cash Money, and Jazzy Jeff get credit for the *transformer*, and Jeff is also one of the first to record the *chirp*.

DJ QBert is one of the most prolific scratch innovators of all time. QBert helped take DJ Flare's innovation, the *flare*, into countless permutations. While QBert doesn't claim to have invented the *crab*, he is certainly the most visible refiner of that technique, as well as countless others.

DJ Babu coined the term "turntablism" in the mid-1990s, around the time that many scratch DJs started calling themselves "turntablists." Some DJs take exception to these terms, but most now consider a turntablist to be a DJ who approaches the turntable as a musical instrument.

Getting Started

Do the stretching exercises in Chapter 14.

Practice every technique with both hands. You'll probably find it easier to do more complex techniques with your dominant hand, but you're eventually going to need both hands to do complex things on the record and the mixer at the same time.

If you're scratching vinyl records, it's easiest to start out using a long sample, like the white noise track on the *Turntable Technique* record, or tracks three and four on side two of a record called *Turntablist's Toolkit*, which are water running in a sink and rushing in a stream. Long, slowly morphing musical tones are also useful.

The advantage in using a long continuous track of noise like those described above is that it won't matter if the needle skips a groove or two, you can just keep on scratching. Once you acquire some basic facility, you'll want to move toward shorter samples and work for accuracy.

The length of the sound you're using is less of an issue when scratching on a CD deck, as there is no needle to skip, and no groove to skip out of. It can still be useful to have a long continuous sound to use when you're getting started, until you get the basic movements down. I've included a couple on the CD that came with the book.

 Track Six

Section One: Faderless Scratches

The follwowing scratches are accomplished entirely by manipulating the record with the faders open. The corresponding CD tracks have examples of each of the scratches, which you can play along with.

Be sure to follow the steps in Chapter 14 to creating a relaxed hand position.

Basic (Baby) Scratch

The basic scratch is also known as the *baby scratch*, and is the basis for many other scratches. With the fader on, you are simply pushing and pulling the record back and forth in rhythm: Forward, reverse, forward, reverse, etc. (Figures 19.1 and 19.2).

Fig. 19.1 and Fig. 19.2. The baby scratch's simple forward and reverse strokes.

Baby scratches are almost always quarter notes or eighth notes.

 Track Seven

The basic scratch as quarter notes and eighth notes.

Drag

When you slow baby scratches down to half notes or whole notes, they become *drags*. You tend to use more of the record when performing drags and the pitch of whatever audio is on the record (the *sample*) is lower. Depending on tempo, quarter note triplets could either be considered drags or baby scratches.

 Track Eight

Drags as half notes, whole notes, and quarter note triplets, which evenly place three notes where there would usually be four.

Scribble

When you speed the baby scratch up to play 16th notes or faster, the same move is called a *scribble*. Since you're moving faster, the pitch of the sample is higher when performing scribbles and you tend to use a smaller portion of the record.

 Track Nine

Sixteenth and thirty-second note scribbles.

Tear

 ### Track Eight

Splitting up the sound of drags or baby scratches by pausing partway through is known as a tear. It's important to come to a dead stop during your pauses in order to put silence between the notes.

One typical pattern is two pushes forward and one pull back, creating triplets:

Forward (pause), forward (pause) reverse

You can do the exact same pattern starting in reverse:

Reverse (pause), reverse (pause) forward

Another pattern is sometimes called a double tear, two pushes forward followed by two pulls back, creating straight eighth or sixteenth notes:

Forward (pause), forward (pause), reverse (pause), reverse (pause)

 ### Track Ten

Quarter and eighth note triplet tears starting forward, followed by quarter and eighth note triplet tears starting in reverse, followed by straight quarter and eighth note double tears.

Practice slowly at first, starting and stopping, starting and stopping. Be sure to practice with both the right hand and left hand.

One-Handed Lazer

This technique got its name since it sounds like a lazer gun from a video game when using an "Ahhh" style sample.

Lazers differ from other non-fader scratches in that the record is flung back and forth with the middle finger. The finger only comes in contact with the record to push it into motion, so the sound is higher (faster) and louder at first. It's important just to graze the record with your finger; too much downward pressure will cause skipping (Figures 19.3–19.5).

Fig. 19.3, Fig. 19.4, and Fig. 19.5. Flinging the record back and forth with one hand.

 ### Track Eleven

One-handed lazers have a different sound than scratches where the hand remains on the record.

Two-Handed Lazers

Using both hands to fling the record around gives you more sonic options and increases your capabilities; you can send the record back and forth faster using both hands. Sending it forward twice, then back (like a tear) is called a zig-zag (Figures 19.6–19.8).

Fig. 19.6, Fig. 19.7, and Fig. 19.8. Using two hands to fling the record..

 Track Twelve

A few two-handed lazer patterns. Once you master these, make up some of your own.

Muted Lazers

Muted lazers (also known as phasers) are accomplished by having one hand applying slight downward pressure to keep the record in place, while the other hand is doing the flinging. The result is more space between the notes (Figures 19.9–19.11).

Fig. 19.9, Fig. 19.10, and Fig. 19.11. Using one hand to hold (or mute) the record while using the other hand to fling.

 Track Thirteen

Muted lazers naturally have more space between notes, and the notes themselves are shorter, more percussive and less flowing than lazers of the non-muted variety.

Swipes

Swipes are a two-handed maneuver that you may want to tackle after you have some experience combining record moves and fader moves.

While one hand is performing a basic scratch, the other hand is grazing the record in a movement much like a fling. The record is grazed (or swiped) in the middle of the stroke of the baby scratch, and it has the effect of momentarily stopping the record, placing silence between two notes (Figures 19.12–19.14).

Fig. 19.12, Fig. 19.13, and Fig. 19.14. Using one hand to swipe (or graze) the record while using the other hand to perform a basic scratch.

 Track Fourteen

Swipes have the effect of turning half notes into quarter notes, and quarter notes into eighths.

Uzi

Uzis are a super-fast back and forth quivering of the record with a single finger (usually the middle finger). The difference between an uzi and a scribble is that the scribble, while fast, is still in time (16th or 32nd notes). The uzi is more like a continuous muscle spasm with your finger on the record.

Fig. 19.15. The Uzi.

At first, you may want to use the band of white noise, as it's likely that the record will skip while you learn to control your spasms (an oxymoron, I know). Eventually, you'll be able to use shorter samples and even sample tips to give your uzis a variety of timbres (Figure 19.15).

 Track Fifteen

An uzi demonstrated with a few different samples.

Section Two: Raising the Bar by Using Shorter Samples

Shorter samples can be found on many scratch records, including those from Thud Rumble and Vital Vinyl. Some vinyl scratch records, like the "Skratchy Seal" record, feature "skipless" tracks, where samples repeat in exactly the same spot every revolution. If the needle skips a groove or two, another instance of the same sample will be located in exactly the same spot.

You should be able to graduate fairly quickly to using shorter samples when scratching CDs, as there are no needles to skip. You'll find a useful selection of shorter samples on the CD that came with this book. These include newly recorded samples that function similarly to the classic samples, which I didn't use for reasons of copyright.

Tracks Three and Four

All of the techniques covered so far should also be practiced using shorter samples. Once you locate a sample you want to use, visualize where it lies on the record or CD dial. You can do this by slowly playing the entire sample back and forth with your hand, noticing where the beginning and end are in relation to the record label or visual display on your CD deck (Figures 19.16 and 19.17).

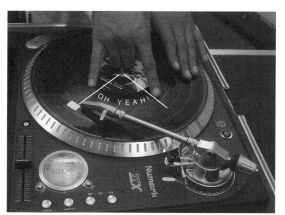

Fig. 19.16 and Fig. 19.17. Visualizing the length of a sample on vinyl and CD.

You can also mark the location of the sample on vinyl, or set a cue point on your CD deck (Figure 19.18).

Sample Tips

One advantage of using shorter samples is your ability to cleanly separate notes with silence by working the initial attack or "tip" of the sound. When using the tips of short samples, forward strokes are accented due to the attack of

Fig. 19.18. Marking an individual sample on vinyl.

the sample. In addition, reverse strokes become shorter; sometimes short enough to be referred to as "staccato" notes.

Locate the beginning of a short sample, then play baby scratch and scribble rhythms with the tip of the sample. Take advantage of the aggressive attack of the sample and the added separation that comes from changing direction during the silence between samples.

 Track Sixteen

Working the tips of samples results in a variety of phrasing variations, including accented long notes and staccato notes.

Combination Tips and Tears

Mixing up tips and tears can result in a variety of phrasing options. Performing tears on shorter samples makes you conserve the space on the record, as you don't want to run out of sound. Visualize where the sample begins and ends, and adjust your strokes to control where you are in the sample.

 Track Seventeen

Patterns that combine tips and tears.

Drum Samples

There are many vinyl records with good drum sounds that you can use for scratching. Look for a song that starts with a drum beat, or has a breakdown section in the track. The Skratchy Seal record and the "Y" record from Thud Rumble offer some good drum samples designed specifically for scratching.

On the CD, a good kick sample comes at the top of track five, followed by more drum samples.

 Track Five

Drum Tips

A popular percussive device in Hip-hop and other styles is to use the tip (or attack) of a drum sample as a fill leading into a downbeat. Kick drums usually work best, but snare drum samples can also be effective.

This sound can often be used while extending breaks, as usually the downbeat of the break is a kick drum. Grandmaster Flash pioneered this usage.

 Track Eighteen

Using drum tips as fills.

Section Three: Scratches Employing the Mixer

Using the mixer's faders and switches in conjunction with record movements adds variety and expression to the attack and release of a multitude of scratches. The scratches in the following section all make use of the mixer in addition to manipulation of the record or CD dial.

If you skipped directly to this chapter, be sure to check out the material on hand and body position in Chapter 14 on approaching the turntable as a musical instrument.

The most commonly used mixer control for scratching is the crossfader, but the channel faders (sometimes referred to as "up faders") and channel switches can also be employed for most of these scratches.

While many DJs start out scratching with their dominant hand manipulating the record, a great deal of them switch to using their dominant hand to manipulate the crossfader once they get into doing crabs, twittles, and flares. I recommend being as ambidextrous as possible.

Basic Fader Scratch

Basic fader scratches are essentially baby scratches where you close the fader each time you change the direction of the record. This gives more separation between each note, and is an easy way to get started combining fader moves and record moves.

Begin with the fader closed:

1. Simultaneously push the record forward and open the fader.

2. Close the fader as you reach the end of your stroke.

3. Simultaneously pull the record back and open the fader.

4. Close the fader as you reach the end of your stroke (Figures 19.19–19.22).

Fig. 19.19, Fig. 19.20, Fig. 19.21, and Fig. 19.22. A basic fader scratch forward and back again.

Track Nineteen

Basic fader scratches in simple rhythms.

Stabs

Stabs are like basic fader scratches, except that you open the fader for the forward stroke only, closing the fader before you change direction and keeping it closed for the pull back.

Begin with the fader closed:

1. Simultaneously push the record forward and open the fader.

2. Close the fader as you reach the end of your stroke.

3. Pull the record back with the fader closed (Figures 19.23–19.25).

Fig. 19.23, Fig. 19.24, and Fig. 19.25. When performing the stab, the fader remains closed for the reverse stroke.

Closely related, "back stabs" simply reverse which part of the stroke the listener hears. The fader is open for the reverse stroke only, and closed for the forward stroke.

 Track Twenty

Stabs and backstabs in various rhythms.

Cutting

Cutting is similar to performing stabs, with one principle difference: Instead of pushing the record forward, you are simply letting it play. Grandmaster Flash pioneered cutting techniques way back in the day.

Cutting is often used with vocal samples, as letting the sample play at regular speed makes it easier to identify what is being said.

 Track Four

Cue up an "Oh, Yeah!" sample or something similar from a track of short samples.

Begin with the fader closed.

1. Hold the record as the platter spins beneath.

2. Simultaneously release the record and open the fader.

3. Close the fader.

4. Pull the record back with the fader closed.

5. Repeat in time (Figures 19.26–19.29).

Fig. 19.26. Release the record and open the fader.

Fig. 19.27. Close the fader.

Fig. 19.28. Pull the record back.

Fig. 19.29. Repeat.

 Track Twenty-Two

Cutting in a variety of rhythms.

Cutting Drums

Sometimes referred to by turntablists as "drum scratching" or "beat scratching" as you are creating a drum beat by manipulating drum samples manually, rather than just letting a drum loop play.

Use a good clean sample of a kick drum right next to a snare, like the ones found on the "Y" record from Thud Rumble, or track five of the CD that came with the book.

Track Five

Start by cutting with the kick only. Cue up the kick sample that plays right before the snare. Mark the sample or notice the exact location on the label or CD display.

With the kick drum cued and the platter spinning underneath:

1. Simultaneously open the fader and play the kick sample.

2. Close the fader.

3. Pull back to the start of the kick sample with the fader closed.

4. Repeat in time.

This is the same as the cutting technique earlier in this chapter; the difference is in the sound.

 Track Twenty-Three

The kick drum, playing first on beats one and three, then a bit busier.

Next, try adding in the snare after every second kick, simply by letting the sample play long enough to include both. It will go something like this:

Kick (pause), kick-snare (pause).

With the kick drum cued and the platter spinning underneath:

1. Simultaneously open the fader and play the kick sample.

2. Close the fader and pull back to the start of the kick sample.

3. Simultaneously open the fader and play the kick and snare sample.

4. Close the fader and pull back to the start of the kick sample.

5. Repeat in time.

 Track Twenty-Four

A simple drum pattern with a kick and snare created by cutting drums.

The tempo in this type of usage is related to the distance between the kick and the snare. Keep in mind that the pitch control may need to be adjusted to accommodate various tempos.

Since there is a second kick following the snare, all you need to do is let the sample play a little longer to include the second kick:

Kick (pause), kick-snare (pause), kick (pause), kick-snare-kick (pause).

 Track Twenty-Five

Live drumming using multiple kick samples before and after the snare.

The above examples are just the beginning. Once you master these, experiment and invent combinations of your own.

A few ideas to try:

1. Isolate the snare and mark it with a different color sticker.

2. Make up patterns that utilize individual snare hits.

3. Experiment playing the sounds backwards.

4. Work samples played backwards as well as forwards into your beats.

5. Work high-hat samples into your beats.

Fades

For this technique, one hand simply fades out the channel fader while the other hand is performing a repetitious record movement such as a scribble or tear. As simple as this sounds, it can be quite effective.

Start with the channel fader all the way up:

1. Play eighth notes as a simple baby scratch.

2. Slowly bring the channel fader down with the other hand.

Once you master the above, try a scribble, a tear, and other more complex patterns. This technique can provide a fitting end to a scratch routine or solo.

 Track Twenty-Six

Fades performed on a variety of record maneuvers.

Uzi Fade (Helicopter)

This is accomplished by combining a fade with uzi scratch. With the right sample (the tip of a kick drum, for instance) this technique sounds like a helicopter flying by.

Cue up your record over the desired sample.

1. Vibrate the record as fast as possible with your middle finger.

2. Slowly raise and lower the channel fader.

Once you get the hang of this, you can play with other controls to give the technique a different effect; panning the helicopter left to right for a fly-by, or tweaking the EQ to increase the intensity (Figure 19.30).

Fig. 19.30. An uzi fade.

 Track Twenty-Seven

A variety of uzi fades, using different samples.

Echoes

When performing echoes, you're manually emulating an echo effect. The advantage is you can play the echo in time without having to set any parameters; you have total control. This technique is a lot like cutting, except you must use the channel (up and down) fader rather than the crossfader. Each time you repeat the sample, you stop the up fader at a slightly lower point than the time before.

If your mixer has lines depicting descending volumes for the channel fader, try using these as visual cues, decreasing the volume of each repeat by one line. Each mixer is different, so you'll need to experiment and adapt.

Many mixers make it possible to adjust the slope of the channel fader. Setting the slope to "slow" will increase the control you have over the fade of your echoes.

 Track Four

Cue up the "Oh, Yeah!" sample from the track of short samples.

Begin with the fader closed, holding the record as the platter spins beneath.

1. Simultaneously release the record and push the channel fader up all the way.

2. Close the fader.

3. Pull the record back with the fader closed.

4. Simultaneously release the record and push the channel fader up almost as high as the previous time.

5. Repeat steps 2–4 (Figures 19.31–19.34).

Fig. 19.31. Release the record, pushing fader all the way up.

Fig. 19.32. Pull the record back, fader closed.

Fig. 19.33 and Fig. 19.34. Each time you cut the sound in, the channel fader peaks at a slightly lower level, giving an echo effect.

 Track Twenty-Eight

Echoes in a variety of rhythms.

Transformer

DJs Spinbad, Cash Money, and Jazzy Jeff developed the transformer in Philadelphia. The technique combines a drag with rapid opening and closing of the fader, slicing the drag into several individual notes.

Begin with the fader closed.

1. Drag the record forward and back over a long sample.
2. Quickly and rhythmically open and close the fader while dragging the record.

The transformer can be performed with the crossfader, channel fader or switch, with differing results. At first, the crossfader may be the easiest to work up to speed, but the channel fader gives more options in terms of dynamics, and the switch gives the most precise attack and release.

 Track Twenty-Nine

Transformers use the opening and closing of the faders or switch to create eighth notes, sixteenth notes, and triplets.

Section Four: Open Fader Scratches

The following scratches begin and end with the crossfader in the center or open position.

Chirp

The chirp was developed by DJ Jazzy Jeff, and can sound like a bird when executed using a sine wave or whistle sample.

 Track Four

Cue up the beginning of the short tone sample (Figures 19.35–19.37).

　　　The chirp has your hands moving first away from each other, then back together. This "flapping" motion has given the chirp the nickname of the "bluebird."

Fig. 19.35. Begin with the fader open, at the attack of a sample.

Fig. 19.36. Simultaneously push the record forward and close the fader, letting the sample sound before the fader closes all the way.

Fig. 19.37. Simultaneously pull the record back and open the fader, letting the sample sound before reaching the tip of the sample.

 Track Twenty-One

The chirp can be played slow or fast, in a variety of rhythms.

　　　Starting and ending the chirp at the beginning (or attack) of a sample makes the execution especially crisp. Chirps are often used to connect different scratches into various combinations.

Reverse or "Hamster" Style

Some west coast DJs developed the practice of plugging their turntables in on the opposite sides of the mixer, effectively reversing the effect of the crossfader. This became known colloquially as "hampster style," after the west coast turntable crew known as the *Bullet Proof Skratch Hamsters*.

　　　For various techniques that employ the crossfader, such as the flare and the crab, playing with the crossfader reversed significantly changes the feel, and consequentially the natural phrasing and overall sound of the technique.

　　　Most mixers now include a reverse or "hamster" switch, which reverses the effect of the crossfader without changing which deck is coming through which channel fader.

　　　Many top turntablists scratch in hamster mode, although most beat juggle with the crossfader in regular mode.

Flare

DJ Flare first introduced the flare in the early 1990s, and it set off a title wave of variations, first by Dailey City DJs, then by DJs from all over the world.

Like the chirp, the flare begins with the fader in the open position. Like the transformer, each forward and reverse stroke of the record is cut into multiple notes using the fader. Like the baby scratch or scribble, changing the direction of the record with the fader open also creates multiple notes.

What's different about the flare is that it requires you to re-calibrate your thinking, so that the default position of the crossfader is "on," and the emphasis is on momentarily stopping the sound (clicking) rather than creating the sound in the first place. It may seem counterintuitive, especially at first. It also sounds really cool, so don't give up.

> **"For my generation, we were always accustomed to starting the scratch with the fader off. Then Flare showed us the flare scratch around 1990. It was starting with the fader in reverse, starting it "on." It was just turning the whole world upside down"—QBert.**

In the following descriptions, "clicking" the fader means bouncing the crossfader closed and then open again (back to the middle) in one quick movement. When the crossfader physically hits one side or the other, you can hear it click; thus the name.

Flares are often described by how many clicks they have. For the "one-click flare," we're splitting each stroke (both forward and reverse) into two notes by placing a click in the middle of the stroke.

Another separation of notes occurs when the direction of the record is changed, causing a slight pause and a change in timbre.

One-click flare

Start with the fader open, at the top of an "Ahhh"-like sample (Figures 19.38–19.41).

Fig. 19.38. Push the record forward, fader open.

Fig. 19.39. Quickly click the fader closed, then open as the record continues.

Fig. 19.40. Reverse direction, pulling the record back with the fader open.

Fig. 19.41. Click the fader closed, then open as the record continues back.

Repeat if desired.

Executing flares with the crossfader in regular mode, as demonstrated in the preceding photos, the thumb is clicking the fader closed while the fingers are pushing the fader back open. Executing flares hamster style (or in reverse mode), the fingers are clicking the fader closed and the thumb is bouncing it back open.

 Track Thirty

A big part of conquering flares is to get the sound into your head. Listen carefully to these examples of one-click flares, and try to emulate their sound, "Whhit-Choo, Whhit-Choo."

The number of notes you hear in a flare is equal to the number of clicks plus one. So the one-click flare gives us two notes during the forward stroke, and two notes in reverse.

It's also important to note that many DJs use the tip of the "Ahhh" or "Fresh" sample to give a more definite attack and separation of notes to that segment of the flare.

Two-click flare

The two-click flare separates each forward or reverse stroke into three distinct notes. The middle note (between the two clicks) is usually the shortest; "Whhit-chit-Choo." When the two-click flare is performed back and forth in the same rhythm ("Whhit-chit-Choo, Whhit-chit-Choo") is also known as the "orbit," a term which can apply to any symmetrical flare scratch.

Start with the fader open, at the top of an "Ahhh"-like sample (Figures 19.42–19.47).

Repeat to orbit.

Fig. 19.42. Push the record forward, fader open.

Fig. 19.43 and Fig. 19.44. Clicking the fader twice, as the record continues forward.

Fig. 19.45. Change direction, pulling the record back, fader open

Fig.19.46 and Fig. 19.47. Clicking the fader twice as the record continues back.

Track Thirty-One

Two-click flares individually and as orbits.

There's an interesting thing that happens to phrasing when you conceive of the fader as being "on" for its default position, and "clicking" the fader quickly closed then open again to separate the notes. The notes are more "legato," meaning that the notes are longer and more connected, with less silence between them.

Three-click flare

The three-click flare is exactly like the two-click flare, except that the three-click flare separates each forward or reverse stroke into four distinct notes rather than three. The middle two notes (in-between the three clicks) are usually the shortest; "Whhit-chi-chi-Choo."

Start with the fader open, at the top of an "Ahhh"-like sample.

1. Push the record forward, fader open.

2. Click the fader three times as the record continues.

3. Reverse direction, pulling the record back with the fader open.

4. Click the fader three times as the record continues back.

Repeat to orbit if desired.

 Track Thirty-Two

Three-click flares in various executions.

As an orbit (repeated), three-click flares set up a rhythm not unlike that of a beat that is prominent in Cajun music.

Section Five: Using Individual Fingers on the Fader

Twiddle

The twiddle is like the transformer, in that the crossfader is being opened and closed quickly to chop a sample up into many notes. However, the twiddle uses the index and middle fingers in succession to do this more rapidly. The thumb acts like a spring, which pushes the crossfader back after each finger has flicked across it.

 Track Six

It may be easiest to use long tones when you are first playing your twiddles and crabs, so you can concentrate on your fader fingers. Eventually, you can combine the twiddle with forward and backward strokes, cutting and other techniques.

The following instructions are for performing the twiddle with the crossfader in regular (non-hamster) mode.

Begin with the thumb applying slight pressure against the crossfader, pushing it closed. This pressure must remain constant, and force the fader closed between each finger flick (Figures 19.48–19.51).

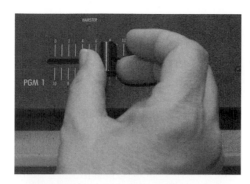

The Twiddle

Fig. 19.48. Flick the fader momentarily open with the middle finger.

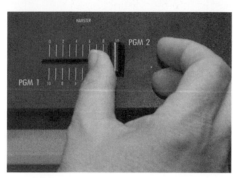

Fig. 19.49. The thumb forces the fader instantly closed.

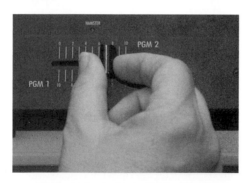

Fig. 19.50. Flick the fader momentarily open with the index finger.

Fig. 19.51. The thumb forces the fader instantly closed.

The above is just one example of a twiddle; you can vary the number of attacks any way you see fit.

Track Thirty-Three

The twiddle makes it easy to play legato 16th notes.

Twiddling with the crossfader in regular mode, as demonstrated in the preceding photos, the fingers are flicking the fader open and the thumb is pushing it back to the closed position.

When twiddling in reverse mode (hamster style), the index and middle fingers are clicking the fader closed and the thumb is bouncing it back open (Figures 19.52–19.55).

   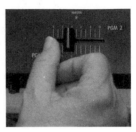

Fig. 19.52, Fig. 19.53, Fig. 19.54, and Fig. 19.55. Twiddling hamster style.

Crab

The crab is similar to the twiddle, in that it uses individual fingers to open and close the crossfader. Instead of two fingers, the crab uses three or four individual fingers in rapid succession across the crossfader to give a rapid-fire series of notes. The other hand is usually playing a sample forward or back, although you can also crab music or samples playing on their own, giving a tremolo effect.

When the crab scratch is performed quickly, the fingers resemble a crab running around on your mixer (if you squint a little bit). The sensation is not unlike clicking your fingers individually in rapid succession against your thumb.

"The concept is simple; you use your thumb to spring," coaches QBert, "and you rub your fingers across, kind of like you're snapping all your fingers."

At first, you may want to play the band of white noise and concentrate entirely on the crossfader. Eventually, you can combine the crab with forward and backward strokes, cutting and other techniques.

The following instructions are for performing the crab with the crossfader in regular (non-hamster) mode.

Four finger crab

Begin with the thumb applying slight pressure against the crossfader, pushing it closed. This pressure must remain constant, and force the fader closed between each finger flick (Figures 19.56–19.60).

The Four-Finger Crab

Fig. 19.56. Flick the fader momentarily open with the pinky.

Fig. 19.57. Flick the fader momentarily open with the ring finger.

Fig. 19.58. Flick the fader momentarily open with the middle finger.

Fig. 19.59. Flick the fader momentarily open with the index finger.

Fig. 19.60. End with the fader closed by the thumb.

Fig. 19.61, Fig. 19.62, Fig. 19.63, Fig. 19.64, and Fig. 19.65. When performing crabs hamster style, the fingers are clicking the fader closed and the thumb is bouncing it back open.

 Track Thirty-Four

Four finger crabs at various tempos (Figures 19.61–19.65).

Eventually, you want the four steps above to happen as one precision motion, like you are snapping your fingers. The best way to learn, and later to improve your crab technique, is to practice slowly at first, and work for accuracy.

"I always start off slow so I can get the intricacies correct, explains DJ QBert, and then I move up to faster speeds as the time goes on."

It's easy to play sloppy crabs, but not very impressive or satisfying for the audience. The masters of this technique perform crabs where each note is distinct, powerful, even, and in time. Crabs are notes, too; often 32nd notes, or 32nd note triplets.

Fig. 19.66. The slope control for the crossfader.

The crossfaders on most professional DJ mixers have an adjustable slope control, which comes in handy when crabbing (Figure 19.66).

Listen carefully as you practice your crabs. If each individual finger isn't getting the crossfader open, make the slope steeper, if you thumb isn't getting the fader closed between each note, make the slope more gradual. Many DJs crab with the slope set as steep as possible, or close to it.

Three finger crabs

Three finger crabs are performed the same as four finger crabs, but you use one less finger, usually leaving out the pinky. You get a three-note pattern rather than a four-note pattern, which is often used as a triplet, or a two 16th notes anticipating a downbeat, which is the third note of the series.

 Track Thirty-Five

Three finger crabs as triplets and straight 16th.

 **Combinations**

Learning the individual scratches is just the beginning. Making up your own combinations is where you start to express your own style. The scratches are like words, they're your vocabulary. Like an Emcee or a lyricist, the DJ puts together sentences and expresses ideas by combining their vocabulary in unique ways. Think about what you want to say through scratching. Don't hesitate to make up some vocabulary of your own!

Additional Resources

For lots of exercises that ease you into scratching, reading music and practicing in a methodical way, check out the book *"Turntable Technique: The Art of the DJ"* from Berklee Press.

On DVD, check out Shortee's *"DJ 101"* and *"202,"* QBert's *"Do It Yourself Volume One: Scratching,"* and yours truly on *"Turntable Technique: The Art of The DJ."*

Index